MW00338044

Investigating
Statistical Concepts,
Applications,
and
Methods

Preliminary Edition

Allan Rossman
California Polytechnic University

Beth Chance
California Polytechnic University

THOMSON

BROOKS/COLE

Australia • Canada • Mexico • Singapore • Spain • United Kingdom • United States

THOMSON

BROOKS/COLE

Acquiring Editor: *Carolyn C. Crockett*
Assistant Editor: *Ann Day*
Editorial Assistant: *Rhonda Letts*
Technology Project Manager: *Burke Taft*
Marketing Manager: *Tom Ziolkowski*
Advertising Project Manager: *Nathaniel Bergson-Michelson*

Project Manager, Editorial Production:
Kelsey McGee
Manufacturing Buyer: *Karen Hunt*
Permissions Editor: *Chelsea Junget*
Cover Designer: *Cuttress and Hambledon/ Hiroko Chastain*
Cover Printing, Printing and Binding:
P. A. Hutchison

COPYRIGHT © 2005 Brooks/Cole, a division of Thomson Learning, Inc. Thomson Learning™ is a trademark used herein under license.

ALL RIGHTS RESERVED. No part of this work covered by the copyright hereon may be reproduced or used in any form or by any means—graphic, electronic, or mechanical, including but not limited to photocopying, recording, taping, Web distribution, information networks, or information storage and retrieval systems—without the written permission of the publisher.

All products used herein are used for identification purpose only and may be trademarks or registered trademarks of their respective owners.

Printed in the United States of America
1 2 3 4 5 6 7   08 07 06 05 04

Brooks/Cole Thomson Learning
10 Davis Drive
Belmont, CA 94002
USA

Asia
Thomson Learning
5 Shenton Way #01-01
UIC Building
Singapore 068808

Australia/New Zealand
Thomson Learning
102 Dodds Street
Southbank, Victoria 3006
Australia

Canada
Nelson
1120 Birchmount Road
Toronto, Ontario M1K 5G4
Canada

Europe/Middle East/Africa
Thomson Learning
High Holborn House
50/51 Bedford Row
London WC1R 4LR
United Kingdom

ISBN 0-534-39110-9

Contents

CHAPTER 1: COMPARISONS AND CONCLUSIONS

In this chapter, you will begin to analyze results from several types of studies. The distinction between different types of studies will be crucial in determining the scope of conclusions that can be drawn from the study results. We will begin focusing on data tables and appropriate numerical and graphical summaries for analyzing the data and effectively presenting the results. Along the way, you will learn some basic terminology and concepts that will be used throughout the course.

CHAPTER 2: COMPARISONS WITH QUANTITATIVE VARIABLES

This chapter parallels the first one in many ways. We will continue to consider studies where the goal is to compare a response variable between two (or more) groups. The difference here is that these studies will involve a *quantitative* response variable rather than a *categorical* one. The methods that we employ to analyze these data will therefore be different, but you will find that the basic concepts and principles that you learned in chapter 1 still apply. These include the principle of starting with numerical and graphical summaries to explore the data, the concept of statistical significance in determining whether the difference in the distribution of the response variable between the two groups is larger than we would reasonably expect from randomization alone, and the importance of considering how the data were collected in determining the scope of conclusions one can draw from the study.

CHAPTER 3: SAMPLING FROM POPULATIONS

The previous two chapters have been about comparing groups. You learned about numerical and graphical summaries of categorical (chapter 1) and quantitative (chapter 2) variables, as well as how to decide whether the difference between two groups is statistically significant. One important caution in interpreting those results was that you often dealt with volunteers or observational data so that you were not be able to generalize beyond that particular group of observational units. In this chapter, you will focus on how to select samples from a larger population so that you may generalize the sample results back to that population.

CHAPTER 4: MODELS AND SAMPLING DISTRIBUTIONS

In the previous chapters, you have developed tools for describing distributions of data and randomization/sampling distributions. In many cases we can provide a simple but more comprehensive description through a probability model. You may not have realized, but you have already seen examples of several models in earlier chapters. In this chapter, we will formalize the notion of a model, see how to check whether our observations conform to a specific model, and then return to the issue of making inferences about a population based on the data in a random sample through these new probability models.

CHAPTER 5: COMPARING TWO POPULATIONS

In chapter 3 you learned about inference procedures for making conclusions about a population parameter or probability when you have randomly selected a sample from that population or process. In chapter 4, you learned about the utility of the normal distribution and the t distribution in modeling the sampling distribution of a statistic, especially when the sample size is reasonably large. In this chapter you will essentially apply these same ideas to the goal of comparing two populations. You will first focus on comparing two different population proportions, where you have randomly selected an independent sample from each population, and then you will focus on comparing two population means, as well as other sample statistics. As you have done before, you will begin by simulating the corresponding sampling distribution and then seeing whether the probability models from the last chapter provide reasonable approximations of these sampling distributions. While we are comparing two groups as we did in chapters 1 and 2, here the randomness arises from drawing the samples from the population, not in the assignment of subjects to groups. Still, you will see that the methods developed here also help us answer the questions we examined in the first two chapters—comparing treatment groups in a randomized experiment. Only in the latter case will we be able to draw any cause and effect conclusions. Only in the former case will we be able to generalize the differences in the samples to differences in the populations.

Section 5-1: Comparing two samples on a categorical response
> Activity 5-1: Newspaper Credibility Decline—Two proportion sampling distribution
> Activity 5-2: Newspaper Credibility Decline (cont.)—Two-sample z-test and z-intervals
> Activity 5-3: Sleepless Drivers—CI for odds ratio

Section 5-2: Randomized experiments revisited
> Activity 5-4: Letrozole and Breast Cancer—Normal approximation for Fisher's Test

Section 5-3: Comparing two samples on a quantitative response
> Activity 5-5: NBA Salaries—Two-sample t procedures (Minitab Exploration)
> Probability Detour
> Activity 5-6: Handedness and Life Expectancy—Effect of sample size and SD
> Activity 5-7: Comparison Shopping (cont.)—Two- versus one-sample with paired data

Section 5-4: Randomized experiments revisited
> Activity 5-8: Sleep Deprivation (cont.)—t approximation to randomization distribution

Section 5-5: Other statistics
> Activity 5-9: Heart transplants and survival—Bootstrap difference in medians and randomization test for medians

hst ylln

CHAPTER 6: COMPARING SEVERAL POPULATIONS, EXPLORING RELATIONSHIPS

The idea of comparing two groups has been a recurring theme throughout this course. In the previous chapters, you have been limited to exploring two groups at a time. You saw that often the same analysis techniques apply whether the data have been collected as independent random samples or from a randomized experiment, although this data collection distinction strongly influences the scope of conclusions that you can draw from the study. You will see a similar pattern in this chapter as you extend your analyses to exploring two or more groups. In particular, you will study a procedure for comparing a categorical response variable across several groups and a procedure for comparing a quantitative response variable across several groups. You will also study the important notion of association between variables, first with categorical variables and then for studies in which both variables are quantitative. In this latter case, you will also learn a new set of numerical and graphical summaries for describing these relationships.

Investigating Statistical Concepts, Applications, and Methods

Preface to the Preliminary Edition

This preliminary edition is intended for use by instructors interested in class testing and providing feedback on these materials. As we continue the revision process, we will place supplementary materials on the Web. These include errata, practice problem solutions, additional homework problems, additional solved examples for reference, sample exams, replacement activities, and a teacher's guide. These will be available for downloading from our Web site (www.rossmanchance.com/iscam/) as we complete them.

To the Student

Statistics is a mathematical science.

While this is a very short sentence, perhaps a self-evident one, and certainly one of the shortest that you will find in this book, we want to draw your attention to several things about it:

- We use the singular *is* and not the plural *are*. It is certainly grammatically correct and more common usage to say "statistics are. . .", but that use of the term refers to statistics as numerical values. In this sentence we mean statistics as a field of study, one that has its own concepts and techniques, and one that can be exciting to study and practice.
- We use *mathematical* as an adjective. Statistics certainly makes use of much mathematics, but it is a separate discipline and not a branch of mathematics. Many, perhaps most, of the concepts and methods in statistics are mathematical in nature, but there are also many that do not involve mathematics. You will see an example of this early in the book as you study the difference between observational studies and controlled experiments. You will find that even in cases where the mathematical aspects of two situations may be identical, the scope of one's conclusions depends crucially on how the data were collected, a statistical rather than a mathematical consideration.
- We use the noun *science*. Statistics is the science of gaining insight from data. Data are (notice the plural here) pieces of information (often but not always numerical) gathered on people or objects or processes. The science of statistics involves all aspects of inquiry about data. Well-designed studies begin with a research question or hypothesis, devise a plan for collecting data to address that issue, proceed to gather the data and analyze them, and then often make inferences about how the findings generalize beyond the particular group being studied. Statistics concerns itself with all phases of this process and therefore encompasses the scientific method.

In these materials, our goal is to introduce you to this practice of statistics, to help you think about the applications of statistics and to study the mathematical underpinnings of the statistical methods. While you will only scratch the surface of the statistical methods used in practice, you will learn fundamental concepts (such as variability, randomness, confidence, and significance) that are an integral part of many statistical analyses. A distinct emphasis will be the focus on how the data are collected and how this determines the scope of conclusions that you can draw from the data.

One of the first features you will notice about these materials is that you will play the active role of investigator. You will read about an actual study and consider the research question, and then we will lead you to discover and apply the appropriate tools for carrying out the analysis. Almost all of the investigations in this book are based on actual scientific studies. At the end of each investigation is a *study conclusion* that allows you to confirm your analysis as well as to see examples of how to properly word conclusions to your studies, for the effective communication of statistical results is as important as the analysis. There are also numerous *explorations* where the primary goal is for you to delve deeper into a particular method or statistical concept. A primary reason for the investigative nature of these materials is that we strongly believe that you will better understand and retain the concepts if you build your own knowledge and are engaged in the context. We don't leave you without support and reference materials: be sure to read the expository passages, especially those appearing in boxes, and the section and chapter summaries. You may find this approach rather frustrating at first, but we also hope you will appreciate developing problem solving skills that will increase in utility as you progress through this and other courses.

We will ask you to use the computer extensively, both to analyze genuine data and also to investigate statistical concepts. Modern data exploration and analysis make heavy use of the computer. We have chosen the statistical package Minitab® as the primary tool for analyzing data. Minitab is increasingly used in industry, but our choice mostly centers on its ease of use. After using Minitab for this course, you will have sufficient background to use most standard statistical packages.

This book will also make heavy use of *simulation* to help you focus on the central question behind many statistical procedures: "How often would this happen in the long run?" Often, the simulation results will direct us to a mathematical model that we can then use as a shortcut; at other times we will only be able to use simulation to obtain an approximation to the answer we are interested in finding. You will make frequent use of several technological tools, namely Minitab, Excel®, and Java™ applets, to carry out these simulations and explorations. We have included instructions for how to use these tools throughout the text so that you may proceed through the investigations with minimal computer and programming background. Still, it will be important that you remember and build on the computing skills that you will develop as the course progresses. All data files and Java applets can be accessed from www.rossmanchance.com/iscam/.

We have also included a series of *practice problems* throughout the book. We envision these exercises as short, initial reviews of the terminology and concepts presented in the preceding investigations. Strategies for learning statistics often mirror those of learning a foreign language —you need to continually practice and refine your use of the terminology and to continually check that your meaning is understood. We hope that you will use these practice problems as a way of quickly assessing your knowledge. Some of the practice problems are in the nature of further explorations, and a few introduce new ideas or techniques based on what you just learned. Not all of the practice problems have simple correct answers but can also be used to spur debate and discussion in your class. Your instructor will inform you about obtaining access to the solutions to these practice problems.

Most of all, we hope you will find fun and engaging examples. Statistics is a vitally important subject, and also fun to study and practice, largely because it brings you into contact with all kinds of interesting questions. You will analyze data from medical studies, legal cases, psychology experiments, sociological studies, and many other contexts. To paraphrase the late statistician John Tukey, "the best thing about statistics is that it allows you to play in everyone else's backyard." You never know what you might learn in a statistics class!

To the Instructor

Motivation/Audience

The statistics education reform movement has revolutionized the teaching of introductory statistics. Key features of this movement include the use of data, activities, and technology to help students understand fundamental concepts and the nature of statistical thinking. Many innovative and effective materials have been developed to support this shift in teaching, and many instructors have changed their approach to teaching introductory statistics. So, where does this book fit in? What problem does it address?

The vast majority of these statistics education reform efforts have been aimed at "Stat 101," the algebra-based service course. As a result, the movement has largely ignored mathematically inclined students, the very students who might be attracted into the field of statistics and into the teaching of statistics. Our goal in this book is to support an introductory course at the post-calculus level around the best features of statistics education reform: data, activities, concepts, and technology.

Mathematically inclined students have typically had to choose between taking Stat 101 or taking a course in probability and mathematical statistics. The first option neither challenges them mathematically nor takes advantage of their mathematical abilities. The second option often devotes an entire course to probability before turning to statistics, which could delay capturing the interest of students who would find data analysis appealing. We offer this book as an alternative that we hope provides a balanced introduction to the discipline of statistics, emphasizing issues of data collection and data analysis as well as statistical inference.

While we believe that this type of introduction to statistics is appropriate for many students, we want to draw particular attention to two student audiences: potential statisticians and teachers. Mathematically inclined students who take Stat 101 may not recognize the mathematical richness of the material, and those who begin with a mathematical statistics course may not appreciate the wide applicability of statistics. We hope that this book reveals the great appeal of statistics by asking students to investigate both the applicability of statistics and also some of its mathematical underpinnings. We also hope to entice students to study statistics further. This book can lay the foundation for several types of follow-up courses, such as a course in regression analysis, design of experiments, mathematical statistics, or probability models.

Prospective teachers of statistics are also an important audience, all the more so due to the growth of the Advanced Placement (AP) program in Statistics and the emphasis on data analysis throughout the K-12 program in the NCTM Standards. Not only is the content of this book in line with the AP course and the NCTM Standards, but so too is the pedagogical approach that emphasizes students' active construction of their knowledge and the use of technology for developing conceptual understanding. We hope that both the content and pedagogy presented here will prepare future teachers to implement similar approaches in their own teaching.

Principles

Some of the principles that have guided the development of these materials are:

- *Motivate with real studies and genuine data.*

Almost all of the investigations in this book center on real studies and genuine data. The contexts come from a variety of scientific disciplines, including some historically important studies. With all of these studies we provide ample background information without overwhelming students or expecting them to know much about the field of application. We also take some examples from popular media, aiming to appeal to diverse student interests. Some investigations ask for data to be collected from students themselves.

- *Emphasize connections among study design, inference technique, and scope of conclusion.*

Issues of study design come up early and recur throughout the book. From the opening chapter we emphasize the distinction between observational studies and controlled experiments, stressing the different types of conclusions that can be drawn from each. We also highlight the importance of randomness and its crucial role in drawing inferences, paying attention to the difference between *randomization* of subjects to treatments and *random sampling* of objects from a population. The connection between study design and inference technique depends heavily on the concept of a sampling/randomization distribution, which we also emphasize throughout.

- *Conduct simulations often.*

We make frequent use of simulations, both as a problem-solving tool and as a pedagogical device. (One challenge with this approach is helping students to recognize the difference between these two uses.) These simulations address the fundamental question underlying many statistical inference procedures: "How often would this happen in the long run?" We often start with tactile simulations before proceeding to technological ones, so that students can better understand the random process being simulated. Technology-based simulations often involve using a Java applet, but we also ask students to write their own small-scale macros in Minitab to conduct simulations. Developing this skill can help students to apply simulation as a general problem-solving tool to other situations.

- *Use variety of computation tools.*

We expect that students will have frequent access to computer software as they work through this book. We ask students to use technology both to analyze data and to explore statistical concepts. Our guiding philosophy is to choose the appropriate software tool for the task at hand. When the task is to analyze data, the appropriate tool is a statistical analysis package. We've chosen Minitab in this book for its ease of use, but other packages could be used as well. When the task is to develop understanding of a concept, the tool is often a Java applet specifically designed for that purpose, typically with a premium on interactivity and visualization. For a few tasks, such as examining the effect of changing a parameter value, the appropriate tool might be a spreadsheet package. We've chosen Excel as the spreadsheet package for this book, but its use is minimal.

- *Investigate mathematical underpinnings.*

The primary contrast between this book and a Stat 101 book is that we often ask students to use their mathematical training to investigate some of the underpinnings behind statistical procedures. An example is that students examine the principle of least squares and other minimization criteria in both univariate and bivariate settings. Students also examine functions symbolically and numerically to investigate issues such as sample size effects. Many of these more mathematical aspects emerge in practice problems after the ideas have been motivated through student investigations of an application.

- *Introduce probability "just in time."*

We don't see probability as the goal of an introductory statistics course, so we introduce probability ideas whenever they are needed to address a statistical issue. Often a probability analysis follows a simulation analysis as a way to obtain exact answers to the simulation's approximation. For example, the hypergeometric distribution is introduced after simulating a randomization test with a 2×2 table, and the binomial distribution arises after using simulation to analyze data from a Bernoulli process. Later probability models are introduced as another type of approximate analysis. Examples include the normal approximation to the binomial for the sampling distribution of a sample proportion and t-distributions as approximations to randomization distributions.

- *Foster active explorations.*

This book consists mostly of investigations that lead students to construct their own knowledge and develop their own understanding of statistical concepts and methods. These investigations contain directed questions that lead students to those discoveries. We expect that this pedagogical approach leads to deeper understanding, better retention, and more interest in the material.

- *Experience the entire statistical process over and over again.*

From the outset we ask students to consider issues of data collection, produce graphical and numerical summaries, consider whether inference procedures apply to the situation, apply inference procedures when appropriate, and communicate their findings in the context of the original research question. This pattern is repeated over and over as students encounter new situations, for example moving from categorical to quantitative responses or from two to several comparison groups. We hope that this frequent repetition helps students to see the see the entire story, to appreciate the big picture of the statistical process, and to develop a feel for "doing statistics." We also emphasize students' development of communication skills so that they can complete the last phase of the statistical process successfully.

Content

Much of the content here is standard for an introductory statistics course, but you will find some less typical inclusions as well. In addition to the early emphasis on study design and scope of conclusions, chapter 1 concentrates on comparisons in the context of categorical response variables, including topics such as relative risk, odds ratio, and Fisher's Exact Test. Concepts introduced here include variability, confounding, randomization, probability, and significance. Then chapter 2 repeats the themes of chapter 1 in the context of quantitative response variables and randomization distributions, also introducing concepts such as resistance. Chapter 3 moves from comparisons to drawing samples from a population, turning again to categorical variables and focusing on hypergeometric and binomial models. Concepts of bias, precision, confidence and types of errors are introduced here. This univariate analysis continues with quantitative variables in chapter 4, where students study more probability models and sampling distributions. They also encounter the t-distribution and a discussion of bootstrapping in this chapter. Chapter 5 then returns to the theme of comparisons between two groups, focusing on large-sample approximations. The final chapter considers comparisons among several groups and association between variables, including chi-square tests, analysis of variance, and simple linear regression.

Pedagogy

This text is designed for use in an active learning environment where students can work collaboratively. We believe that the ideal classroom environment provides constant computer availability for students, but the materials are flexible enough that an instructor can use the investigations as examples through which to lead students in a lecture setting without a computer lab. The study conclusion and discussion sections provide exposition to help students make sure that they are discovering what they are expected to in the investigations.

Structure

Most of this book consists of investigations that ask students to discover and apply statistical concepts and methods needed for analyzing data gathered to address a particular research question. The investigations contain a series of directed questions, with space provided for students to record their responses. At the end of an investigation is a study conclusion that summarizes what the student should have discovered about the study, often followed by a discussion of the statistical issues that emerged. Interspersed among the investigations are explorations that ask students to delve deeper into a statistical concept, often involving the use of technology. We also provide practice problems that assess students' level of understanding of what the preceding investigations were meant to convey. A final component that you will find are detours that introduce terminology or technology hints; we set these apart so as not to interrupt the flow of the investigation of a study.

Pre-requisites

This book provides an introduction to statistics for students who have completed a course in one-variable calculus. We do not make frequent use of calculus, but we do assume that students are comfortable with basic mathematical ideas such as functions, and we do call on them to use derivatives and integrals on occasion. We do ask students to use technology heavily, including some small-scale programming, but we do not assume prior knowledge of programming ideas or of any particular software. No prior knowledge of statistics is assumed, although we do not devote much time to ideas (such as mean and median, histograms, and scatterplots) that they are likely to have encountered before.

Acknowledgements
We thank the National Science Foundation for supporting the development of these curricular materials through grants #9950746 and #0321973.

We are very grateful to colleagues who have class tested or reviewed drafts of these materials: Ulric Lund and Karen McGaughey of Cal Poly, Robin Lock of St. Lawrence University, Jackie Dietz of Meredith College, John Holcomb of Cleveland State University, Chris Franklin of the University of Georgia, Julie Legler of St. Olaf College, and Julie Clark of Hollins University.

We especially appreciate the help that we have received from students who have reviewed the materials and helped with a variety of tasks related to this project. Many thanks to Laurel Koester, Tierra Stimson, Nicole Walterman, and Rebecca Russ. Special thanks to Carol Erickson for always going above and beyond to support our work.

Most of all, we thank all of our students who have served as our pedagogical guinea pigs over the years, for their patience and good cheer in helping us to class test drafts of these materials, and for inspiring us to strive constantly to become better teachers.

CHAPTER 1: COMPARISONS AND CONCLUSIONS

In this chapter, you will begin to analyze results from several types of studies. The distinction between different types of studies will be crucial in determining the scope of conclusions that can be drawn from the study results. First you will focus on data tables and appropriate numerical and graphical summaries for analyzing the data and effectively presenting the results. Along the way, you will learn some basic terminology and concepts that will be used throughout the book.

SECTION 1-1: SUMMARIZING CATEGORICAL DATA

In this this investigation you will be introduced to many basic terms and techniques which you will practice applying throughout the chapter.

Investigation 1-1: Popcorn Production and Lung Disease

In May 2000, eight people who had worked at the same microwave popcorn production plant reported to the Missouri Department of Health with fixed obstructive lung disease. These workers had become ill between 1993 to 2000, while employed at the plant. On the basis of these cases, researchers began conducting medical examinations and environmental surveys of workers employed at the plant in November 2000 to assess their occupational exposures to certain compounds (Kreiss et al., 2002). As part of the study, current employees at the plant underwent spirometric testing. This measures FVC—forced vital capacity—the volume of air that can be maximally, forcefully exhaled. On this test, 31 employees had abnormal results, including 21 with airway obstruction.

A key focus of this study is whether or not an employee had airway obstruction. We will now examine different techniques for organizing and describing such data, focusing on graphical and numerical summaries of the data and how to compare different groups.

(a) It is difficult to judge whether 21 employees with airway obstruction is a lot until we at least know how many employees were tested. Since they tested 116 employees, what *proportion* of these employees had airway obstruction?

$$\frac{21}{116}$$

For *proportions*, always report a decimal value between 0 and 1, inclusive. This proportion can also be referred to as the *baseline rate* or *risk* of airway obstruction in this plant.

Researchers were curious as to whether the baseline rate was high among microwave-popcorn workers. We could compare these workers to the general population (Chapter 3) and we can also compare these workers to others within the same plant who had different levels of exposure. The plant itself was broken into several areas such as the flavor-mixing room, a quality-control room for popping sample product, and packaging rooms. Some of the areas, such as the bag-printing area and the outdoors, were separate from the popcorn production area. Air and dust samples in each job area were measured to determine the exposure to diacetyl, a marker of organic-chemical exposure. The cumulative exposure for each participant was determined by taking into account how long they spent at different jobs within the plant and the average exposure in that job area. Employees were then classified into two groups: a "low exposure group" and a "high exposure group."

These researchers were interested in whether the employees in the high exposure group had a higher "risk" of developing lung problems than those in the low exposure group. If we morbidly define a "success" to be "having airway obstruction," this is the same as comparing the "proportion of successes" in the two groups.

Of the 21 participants with airway obstruction, 6 were from the low exposure group and 15 were from the high exposure group.

(b) While it is clear that there is a higher *number* of high exposure employees with airway obstruction than low exposure employees, suggest a better way to compare these two groups.

In this case, the two exposure groups contained the same number of employees (58). However, if these values had differed, comparing the counts alone would not be very informative. Thus, we will always tend to work with the proportions instead of the counts.

Let's organize the information we have so far. When we want to compare "success" and "failure" outcomes between groups, this is typically done with a "two-way" frequency table.

(c) Fill in the appropriate cells of the following table:

	low exposure	high exposure	Total
Airway obstructed	6	15	21
Airway not obstructed	52	43	95
Total	58	58	116

The first step in data analysis is to **look at your data!** You will always have choices in how you display the data. We will try to highlight what are considered the most effective displays depending on the type of data you are working with. The second step will be to provide some suitable numerical summaries of the information. Again, you will often have some choices.

GRAPHICAL SUMMARY

> **Def:** A *segmented bar graph* is a visual display that compares a categorical variable across two groups that presents the *proportion* or *percentage* of times that each outcome occurs in each group. Using the percentages is much more informative than only indicating the *number* of occurrences, especially if the groups have different sample sizes. For example, see Figure 1 below.

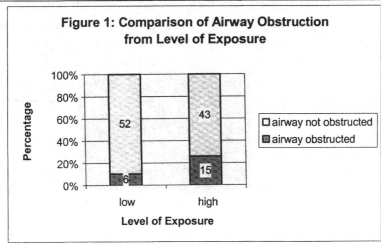

(d) Recreate this graph using Excel:
- Enter the two-way table above into cells A1:C3 (include row and column headings but not the row or column totals):

	A	B	C
1		low exposure	high exposure
2	airway obstructed	6	15
3	airway not obstructed	52	43

- Drag your mouse over these 9 cells to highlight them.
- Open the Chart Wizard (Insert > Chart) and use the "Column" Chart type. For the chart sub-type, select the "100% stacked column" option.
- Click the "Next" button twice, and then add a chart title and axis labels. You can also select the Data Labels tab and choose the "Value" option. Then click "Finish".
- If you double click on the gray background in the chart, you can change the area coloring to "None".

(e) Describe what this graph tells you about how the two groups compare. *Habit*: Try to describe the graph to someone who does not have a first hand view of it, and always write in terms of the context of the study.

NUMERICAL SUMMARIES

(f) Calculate the proportion (a decimal value between 0 and 1) of employees in each exposure group who exhibited airway obstruction. Round your answers to 3 decimal places.

Def: Since we calculated the proportion of successes separately for each group, "conditioning" on the explanatory variable group, these are often referred to as *conditional proportions*. If these conditional proportions differ across groups, we often say there is an *association* or *relationship*.

In this study, the conditional proportion of employees with airway obstruction in the high exposure group is larger than the conditional proportion of employees with airway obstruction in the low exposure group. Since these conditional proportions (where we have *conditioned* on the level of exposure) differ, there appears to be an *association* between level of exposure and development of airway obstruction.

(g) One way to compare the two groups is to look at the difference in these two proportions (or *risks*). Subtract the proportion of the low exposure group with airway obstruction from the proportion of the high exposure group with airway obstruction. Does this seem like a large difference to you?

(h) Suppose the two proportions had been .650 and .494. What is the difference in this case? Does this seem like a large difference to you? Explain.

The difference in the proportions (that is risks) is the same, but for our data the risk of airway obstruction in the high exposure group is more than twice as big as in the low exposure group. For this reason, the difference alone may not be the best way to compare these two proportions, especially when they are far from .5.

Def: Another way to numerically compare two groups is the *relative risk*, the ratio of the risks for each group.

$$relative\ risk = \frac{risk\ of\ group\ 1}{risk\ of\ group\ 2} \quad \text{or} \quad \frac{proportion\ of\ successes\ in\ group\ 1}{proportion\ of\ successes\ in\ group\ 2}$$

Note: it is often standard to put the group with the lower risk in the denominator.

(i) Calculate the relative risk for the table you constructed in (c).

Discussion: We interpret this number by saying that the risk of airway obstruction is 2.5 times greater for employees in the high exposure group than for employees in the low exposure group. In reporting relative risk, you should include information about the *sample size* (total number of observations) of the study and the overall baseline risk. Another numerical comparison between the groups discusses the data in terms of *odds*.

Def: The *odds of success* is defined as the ratio of the proportion of "successes" to the proportion of "failures" which simplifies to the ratio of the number of successes to failures.

$$odds = \frac{proportion\ of\ successes\ in\ the\ group}{proportion\ of\ failures\ in\ the\ group} = \frac{number\ of\ successes\ in\ the\ group}{number\ of\ failures\ in\ the\ group}$$

For example, if the odds are 2-to-1 in favor of an outcome, we expect a success twice as often as a failure in the long run. If the odds are 9-to-5 in favor of an outcome, we expect roughly 9 successes for every 5 failures in the long run. It's important to note how the "outcome" is defined. For example, in horse racing, odds are typically presented in terms of "losing the race," so if a horse is given 2-to-1 odds against winning a race, we expect the horse to lose two-thirds of the races in the long run.

(j) If we treat "having airway obstruction" as a success, calculate the overall odds of airway obstruction for these 116 workers:

> **Def:** To compare the odds of successes for two different groups, we compute the *odds ratio* between group 1 and group 2 (using group 2 as the "reference" group):
>
> $$odds\ ratio = \frac{odds\ of\ success\ in\ group\ 1}{odds\ of\ success\ in\ group\ 2}$$
>
> Note, it is often standard to put the group with the lower odds in the denominator so that the odds ratio is larger than 1.

(k) Compute the odds ratio of airway obstruction between the high exposure group and the low exposure group:

      Odds of airway obstruction for high exposure group:

      Odds of airway obstruction for low exposure group:

      Odds ratio (high exposure over low exposure):

Discussion: You should have found that the odds of airway obstruction for the high exposure group is 15/43, or roughly 1-to-3, and the odds of airway obstruction for the low exposure group is 6/52, roughly 1-to-9. We interpret the ratio by saying that the odds of airway obstruction in the high exposure group are estimated to be 3.02 times larger than the odds of airway obstruction in the low exposure group.

Cautions: Before we finish describing our data, it is always important to consider some cautions in the interpretations.
- This was a "cross-sectional" design. The researchers only gathered information about the current status of current workers and the current plant conditions at one point in time. It is important to remember that the measured exposures may not reflect historical exposures. This study did not track other sources and levels of exposure for these workers. This is an example of the broad classes of studies generally referred to as *observational studies.*
- There may also be a "healthy worker effect." In this plant, eight former workers were known to have left their jobs because of lung disease, thus leaving behind a healthier workforce to participate in the study. It is even important to consider which workers were absent (perhaps home sick?) when data about the workers' health were being recorded.

Study Conclusions: Employees with high exposure to diacetyl (sample size 58) were two and half times more likely to experience airway obstruction (or, the odds were 3.03 times higher) than those in the low exposure group (sample size 58). While there is evidence of a difference in the risk airway obstruction between the two groups, we cannot draw a cause-and-effect conclusion between the level of exposure and the risk of airway obstruction from this study nor generalize these findings to all workers in the factory because of the cautions suggested above. Later in this chapter you will see how to design a study so cause-and-effect conclusions are valid, as well as how to decide when a difference between groups is large enough to convince us that is represents a genuine difference between the groups. In Chapter 3, you will consider whether the study results can be generalized beyond the individuals involved in the study.

Terminology Detour: To improve our communication about data, we need to define several terms that will be used throughout the book.

> **Def:** The *observational units* are the people or objects about which data are recorded. A *variable* is any characteristic that varies from observational unit to observational unit.

> In the previous example, the observational units are the employees. The forced-vital-capacity measurements, exposure level, age, and gender, are all variables that we could measure about these employees. However, "number of employees" is not a variable for these employees since it summarizes information about the groups instead of recording a characteristic that varies from employee to employee. If we were to treat the observational units as different manufacturing plants rather than employees, then "number of male employees" would be a legitimate variable across the plants. It is often helpful to define the variable as a question that could be posed to each observational unit. For example: How old is the subject? Did the subject have airway obstruction? How many male employees does each plant have?

> **Def:** We classify the type of variables as *categorical* (assigning each observational unit to a category) or *quantitative* (assigning each observational unit a numerical measurement). A special type of categorical variable is a *binary* variable which has just two possible outcomes.

> One way to help decide between these types is to ask yourself whether or not you can take the average of the values. For example, gender is categorical but the FVC (volume of air) measurement is quantitative. The zip code of the employee's home address would be categorical even though it is a number, because arithmetic operations like adding or averaging do not make sense on zip codes. Age can often be treated either way: we will consider it quantitative if the data were recorded in such a way that it makes sense to discuss the average age of the employees, but not if they have been grouped into age categories. Often quantitative variables can be turned into categorical variables; instead of using the quantitative FVC measurements, we focused on the categorical "did they have airway obstruction" variable.

> **Def:** We often classify variables as *explanatory* and *response*, depending on their role in the study. Response variables are the measured or observed outcomes of interest and the explanatory variables are potential explanations for any changes in the response.

> For example, if we want to investigate whether "studying improves your grades," then the amount of time you spend studying is an explanatory variable and your grade is a response variable. In the popcorn study, since the researchers were interested in whether employees in the high exposure group had a higher "risk" of developing lung problems than those in the low exposure group, the explanatory variable is the level of exposure (categorical) and the response variable is whether or not the employee developed lung problems (categorical). You will also see the response variable referred to as the *dependent variable* or *outcome* and the explanatory variable as the *independent variable* or *predictor variable* or *exposure*.

> **Def:** The *distribution* of a variable describes how the variable behaves, that is, the possible outcomes of the variable and how often each outcome occurs relative to the other outcomes.

Conventions: In constructing a two-way table, it is standard to use the explanatory variable as the column variable and the response variable as the row variable. When constructing a segmented bar graph, we put the explanatory variable categories along the horizontal axis. In calculating the conditional proportions, we typically condition on the categories explanatory variable, for example "of those in the high risk group, what proportion had airway obstruction."

SECTION 1-1 SUMMARY

A first step in data analysis is to identify the *observational units* in the study and the *variable(s)* measured on these units. In this section, you focused on comparing the *distribution* of *response variable* outcomes between two or more categories of the *explanatory variable*. As you will see numerous times in this text, we will describe these distributions through graphical displays and numerical summaries. When dealing with a *binary* response variable, we focus on the proportion of *successes* and *failures*. In this setting, an effective graphical display is a *segmented bar graph* (easily created in Excel). You examined several different numerical summaries including the *difference in conditional proportions, relative risk*, and *odds ratio* for comparing the response between two explanatory variable groups. The first summary has some limitations as it does not provide an indication of the size of the difference relative to the baseline rate (the overall proportion of successes in the entire collection of observational units). Relative risk and odds ratio provide such information, but with slightly different interpretations (the ratio of the risks versus the ratio of the odds) and typically compare one explanatory variable group to a reference or "control" group. As you continue to calculate relative risks and odds ratios in the next section, you should pay attention to how "success" has been defined in the study, as well as to which variable is being considered the explanatory variable and which the response variable. In this section you also saw that after considering the information conveyed by the numerical and graphical summaries, you still need to be cautious in how broadly you try to generalize the conclusions. This will be a theme throughout this text.

Practice Problems: You will notice a plethora of "practice problems" in this text. The intention is to give you an opportunity to apply the most recently covered material and provide you with an opportunity to use the recently introduced terminology. Not all of the practice problems have clear answers and may instead be used to initiate discussions and debate. Occasionally, a practice problem will ask you to expand your learning a bit beyond what has been presented in the section. We encourage you to worry less about completely correct answers here and more with determining where you want to ask additional questions.

1-1) Go take a hike!
The book *Day Hikes in San Luis Obispo County* by Robert Stone gives information on 72 different hikes that one can take in the county. For each of the 72 hikes, Stone reports the distance of the hike (in miles), the anticipated hiking time (in minutes), the elevation gain (in feet), and the region of the county in which the hike can be found (North County, South County, Morro Bay, and so on, for a total of eight regions).
(a) What are the observational units here?
(b) How many variables are mentioned above?
(c) Classify each variable as quantitative or categorical.
(d) If you create a new variable that is the ratio of hiking time to hiking distance, would that be a quantitative or categorical variable?
(e) If you create a new variable called "length of hike" that is coded as "short" for a hike whose distance is less than 2 miles, "medium" for a hike whose distance is at least two but not more than four miles, and "long" for a hike whose distance is at least four miles, would this be a quantitative or categorical variable?
(f) For each of the following, specify whether or not it is a legitimate variable for the observational units you specified in (a).
- Longest hike in the book
- Average elevation gain of a hike in the book
- Whether or not the hike is in the North County region
- Proportion of hikes with an elevation gain of more than 500 feet
- Anticipated hiking time, reported in hours

1-2) Identifying variables
(a) Consider the 50 states as the observational units for a study. Suggest one quantitative and one categorical variable that you could measure about the states. (Be careful to express your answers clearly as variables.)
(b) Consider the 9 members of the Supreme Court as the observational units for a study. Suggest one quantitative and one categorical variable that you could measure about the justices. (Be careful to express your answers clearly as variables.)

1-3) More popcorn problems
Another part of the popcorn production study classified current workers according to where they worked. Twenty employees working on the plain-popcorn packaging line, the bag-printing areas, the warehouse, the offices, and the outdoor areas were considered an internal reference group to the employees who worked in the other work areas. Trained interviewers administered a standardized questionnaire to employees of the popcorn plant (written informed consent was obtained from all participating employees). One of the questions on the survey concerned "exertional shortness of breath" (shortness of breath when hurrying on level ground or walking up a slight hill), which was reported by 31 participants, 1 from the reference group of 20 employees and 30 from 92 of the production employees.
(a) Identify the explanatory and response variables. Specify whether they are quantitative or categorical.
(b) Construct the two-way table and segmented bar graph comparing the exertional shortness of breath of the reference group to the production employees.
(c) Calculate the baseline rate of exertional shortness of breath, the relative risk, and the odds ratio to compare the exertional shortness of breath of the reference group and the production employees. Include interpretations of these numbers.
(d) Do the production employees appear more likely to have exertional shortness of breath than the reference group? Give reasons to support your answer.
(e) Not all of the workers in the plant completed the questionnaire. Suggest some cautions you might have in interpreting the results because of this.

SECTION 1-2: ANALYZING CATEGORICAL DATA

In this section, you will practice applying the terminology from the previous section and you will think more carefully about the types of conclusions that can be drawn from such studies. You will also continue to investigate properties of relative risk and odds ratio.

Investigation 1-2: Smoking and Lung Cancer

After World War II, evidence began mounting that there was a link between cigarette smoking and pulmonary carcinoma (lung cancer). In the 1950s, two now classic articles were published on the subject. One of these studies was conducted in the United States by Wynder and Graham ("Tobacco Smoking as a Possible Etiologic Factor in Bronchiogenic Cancer", *Journal of the American Medical Association*). They found records from a large number (684) of patients with proven bronchogenic carcinoma (a specific form of lung cancer) in hospitals in California, Colorado, Missouri, New Jersey, New York, Ohio, Pennsylvania, and Utah. They personally interviewed 634 of the subjects to identify their smoking habits, occupation, exposure to dust and fumes, alcohol intake, education, and cause of death of parents and siblings. Thirty-three subjects completed mailed questionnaires and information for the other 17 was obtained from family members or close acquaintances. Of those in the study, the researchers focused on 605 male patients with the same form of lung cancer. Another 1332 hospital patients with similar age and economic distribution (including 780 males) without lung cancer were interviewed by these researchers in St. Louis and by other researchers in Boston, Cleveland, and Hines, IL.

The following two-way table replicates the counts for the 605 male patients with the same form of cancer, and for the "control group" of 780 males.

Amount of Cigarette Smoking for more than 20 years by males*

	None (less than 1 per day)	Light (1-9 per day)	Moderately Heavy (10-15 per day)	Heavy (16-20 per day)	Excessive (21-34 per day)	Chain (35+ per day)
Lung Cancer Patients (*n*=605)	8	14	61	213	187	122
Controls (*n*=780)	114	90	148	278	90	60

* Pipe and cigar smoking were included by counting 1 cigar as 5 cigarettes and 1 pipeful as 2½ cigarettes.

(a) What are the observational units in this study?

(b) Identify the explanatory variable and the response variable for this study. Identify each as quantitative or categorical. If categorical, how many categories are there?

       Explanatory variable: type:

       Response variable: type:

(c) Use Excel to construct a segmented bar graph. Typically we use the explanatory variable along the horizontal axis. [If necessary, to change which variable appears on the horizontal axis, after you create a graph, if you right click on a bar and choose "Source data," you can specify the "Series in:" option to be columns instead of rows.] Describe how these groups compare.

(d) Consider only the 226 men who were non-smokers or light smokers.

	Non-smoker (less than 1 per day)	Light smoker (1-9 per day)
Lung Cancer Patients (*n*=605)	8	14
Controls (*n*=780)	114	90

If we define "having lung cancer" as a success, compute the odds ratio of having lung cancer for the light smokers compared to the non-smokers.

Odds of lung cancer (light smokers):

Odds of lung cancer (non-smokers):

Odds ratio:

(e) Simplify the above calculation to express this odds ratio as a cross product involving the four entries in the table.

Short-cut: For a general 2×2 two-way table, the odds ratio can be calculated as

a	b
c	d

$\dfrac{ad}{bc}$ or $\dfrac{bc}{ad}$ such that the odds ratio is at least 1 or depending on the choice of reference group.

(f) Compute the odds ratio of lung cancer for the "heavy" smokers compared to the non-smokers. *Hint*: Start by extracting the relevant two-way table from the full table and apply the short-cut:

	None (less than 1 per day)	Heavy (16-20 per day)
Lung Cancer Patients (*n*=605)		
Controls (*n*=780)		

      Odds ratio:

(g) Compute the odds ratio of lung cancer for the "chain" smokers compared to the non-smokers and interpret this value.

      Odds ratio:

      Interpretation:

(h) Compute the odds ratio of lung cancer for smokers (any amount) compared to non-smokers. Interpret this value.

(i) Do these odds ratios reveal a relationship between smoking and lung cancer? Describe the relationship.

STUDY CONCLUSIONS
(j) When this study was published, there were numerous criticisms of its ability to establish a *causal* connection between smoking and lung cancer. What do you think some of these criticisms were?

(k) The researchers first compared the age distributions of the lung cancer group and the control group. Why do you think this was important?

(l) From these data, would it be reasonable to conclude that roughly 605/(605+780), or 44%, of *all males* of similar ages and economic status, are smokers? Explain.

Study Conclusions, I: This study reveals that the odds of lung cancer are about 13 times higher for smokers than for a comparable group of nonsmokers (matched, for example, by age and economic status). The data also show that the odds of having lung cancer increase with the amount of smoking (light smokers have 2 times the odds, heavy smokers have 11 times the odds, and chain smokers have 29 times the odds!)—this is called a "dose-response." There is a strong relationship between the size of the "dose" of smoking and occurrence of lung cancer for these patients.

So far, these conclusions only apply to the individuals in this study, as in the popcorn study, and while the data suggest an association, this is quite different from saying "smoking causes lung cancer." We also may not be willing or able to believe that the distributions observed in this study apply to a larger group of observational units as well. In determining the scope of conclusions that you can draw from a study, you will always have to consider how the data were collected.

Terminology Detour: In describing methods for data collection, we will consider both how the subjects in the study were selected and how the explanatory variable groups were determined. The following terminology will help this discussion and will be used throughout the text.

Def: The *sample* is the collection of observational units on which the variables have been measured. Often, we will want to think of this sample as a subset of a larger *population* of interest. Ideally, we would like to take what we learn from the sample and generalize the conclusions to the larger population, but we must carefully consider how the sample was selected from the larger population.

In the above example, the sample was the 1385 male hospital patients. The population could be considered all male patients of similar age and economic status.

Def: In an *observational study*, the researchers record the values of the variables of interest and make no attempt to influence the outcome of any of these variables. There are several types of observational studies, including:

- *Case-control study.* The researchers identify observational units in each response variable category (the "cases" and the "controls") and then determine the explanatory variable outcome for each observational unit. How the controls are selected is very important in determining the comparability of the groups. These are often *retrospective designs* in that the researchers may need to "look back" at historical data on the observational units.

- *Cohort study.* The researchers select individuals according to the explanatory variable and then observe the subsequent outcomes of the response variable. These are usually *prospective designs* and may even follow the subjects for several years.

- *Cross-classification study.* The researchers categorize subjects according to both the explanatory and the response variable simultaneously. For example, they could take a sample of adult males and simultaneously records both their smoking status and whether they have lung cancer. A common design is *cross-sectional*, where all observations are taken at a fixed point in time.

The Wynder-Graham study was a *case-control study*. The researchers identified subjects in each response category (the cases having lung cancer and the controls not), and then determined whether or not they smoked. This was also a *retrospective design* in that the researchers looked into the past to determine the explanatory variable outcome. The microwave popcorn plant study was a cross-sectional, cross-classification design as the exposure levels and airway obstruction were measured for each observational unit. Though it is important to keep in mind that the researchers probably defined "high" and "low" exposure to create groups of similar sizes.

Cause and Effect Conclusions
(m) Would it be reasonable to draw a cause and effect conclusion from a case-control study? Explain.

Generalizing from Sample to Population
(n) Would it be reasonable to generalize the relationship between smoking and lung cancer observed in this sample to the population of all males of similar ages and economic status? Explain.

Discussion:
While case-control designs allow for immediate collection of data and ensure a reasonable group size in each response variable category, there are limitations in the conclusions we can draw from such a study.

Cause and Effect Conclusions
   One limitation of this study is we don't know for sure if the control group of patients without lung cancer truly "matches" all relevant characteristics of the disease group. For this reason, we cannot draw a cause and effect conclusion between smoking and lung cancer despite the strong association observed. For example, there could be some other difference between the smokers and non-smokers, such as a genetic characteristic, that is actually the link to development of lung cancer.

Generalizing from Sample to Population
   Another limitation is that we don't know whether this sample of hospitalized patients is representative of the population of all American males. In fact, we are pretty sure that the rate of lung cancer in this sample does not mirror the population rate because the researchers actively controlled that variable to ensure a large number of patients in each group. Thus we have no information about how frequently lung cancer occurs naturally in the population. [In fact, some people may be hesitant to refer to "whether or not they have lung cancer" as a *variable* since each person's result was predetermined by the design of the study.] In conclusion, we cannot generalize our finding of an association between smoking and lung cancer to the larger population.

Measurement Issues
   Furthermore, these types of studies are also subject to "retrospective bias," relying on the subjects to accurately remember, and being willing to tell, details of their life styles. This can also be complicated by asking these questions of patients who know they have been diagnosed with lung cancer, as their recall may be affected by this knowledge.

ODDS RATIO VS. RELATIVE RISK EXPLORATION

Suppose we did want to make some statements about the population of adult males. In Chapter 3, you will learn techniques for selecting the sample that make such generalizations reasonable. However, sometimes the study design and numerical summary involved also affect these generalizations. You considered above that it was not reasonable to interpret the base rate of lung cancer from these sample data since the researchers controlled the number of lung cancer and non-lung cancer patients present in the study. That is, when the distribution of the response variable was controlled by the study design instead of observed naturally, we will not interpret the baseline rate.

> When the distribution of the response variable is controlled as part of the study design, it is not appropriate to generalize the baseline rate of success from the sample to the population. By the same logic, it is not appropriate to interpret the conditional proportion of successes from either explanatory variable group. Consequently, it is not appropriate to generalize the relative risk to the general population in case-control studies. If you have a cohort or a cross-classified design, interpretations of relative risk are appropriate. However with a cohort study, you would not want to generalize the risk of "success" for the different explanatory variable groups. The key observation is whether or not the distribution of the variable of interest is controlled by the researcher.

In the next set of questions, you will compare how the relative risk and odds ratio calculations depend on which variable has been identified as the response variable and which category has been identified as success. Return to the two-way table of the 226 non-smokers and smokers.

	Non-smoker	Light smoker	Total
Lung cancer	8	14	22
Control	114	90	204
Total	122	104	226

(o) Consider being in the control group as a success and compute the odds ratio of being in the control group for the non-smokers compared to the light smokers. Show the details of this calculation directly instead of using the short-cut calculation. How does your answer to compare to your answer in (d)?

Odds of control group (non-smokers):

Odds of control group (light smokers):

Odds ratio:

(p) Consider the smoking variable as the response variable and being a "non-smoker" as a success (and light smoker as a failure). Compute the odds ratio of being a non-smoker for the control group compared to the lung cancer patients. How does your answer to compare to your answer in (d)?

Odds of non-smoker (control group):

Odds of non-smoker (lung cancer group):

Odds ratio:

(r) Compare this to the relative risk of "lung cancer" for the light smokers compared to the non-smokers.

Risk of lung cancer (light smokers):

Risk of lung cancer (non-smokers):

Relative risk:

(q) Calculate the relative risk of "no lung cancer" for the non-smokers compared to the light smokers.

Risk of no lung cancer (non-smokers):

Risk of no lung cancer (light smokers):

Relative risk:

(s) Compare this to the relative risk of "non-smoker" for the "no lung cancer" group compared to the "lung cancer" group.

Risk of non-smoker (lung cancer group):

Risk of non-smoker (no lung cancer group):

Relative risk:

(t) Did the odds ratio change when you switched the roles of the explanatory variable and the response variable? Did the odds ratio change when you switched which outcome was considered success and which failure? What about the relative risk calculations?

Discussion: The above calculations have shown that the odds ratio calculation is *invariant* to (does not depend on) the choice of explanatory vs. response variable and the choice of success and failure, as you found an odds ratio of (14×114)/(8×90) = 2.217 in all three cases you examined. However, the relative risk calculation is not invariant (the correct values are 2.05, 1.08, and 1.44). In particular, our conclusion about the strength of the relationship depends on which calculation we examine (the relative risk of "no lung cancer" for non-smokers does not appear as extreme as the relative risk of "lung cancer" for the smokers). However, the odds ratio calculation does not depend on which variable is treated as the explanatory and which as the response (the odds ratio of lung cancer for non-smokers is the same as the odds ratio of smoking for lung cancer patients). This is relevant for case-control studies. We would like to describe the risk of lung cancer from smoking, but cannot interpret the relative risk from case-control studies. Yet the odds ratio calculation is the same whether we consider smoking the response or explanatory variable, and similarly the interpretation of the odds ratio calculation does not change depending on whether a variable was controlled or observed. So it is meaningful to interpret the odds ratio for any type of study (case-control, cohort, or cross-classified). For this reason, we will tend to use the odds ratio as a measure of association between two binary variables instead of relative risk, even though relative risk is easier to interpret. Fortunately, when the conditional proportions of success are small, as they often are in medical studies for example, the values of the odds ratio and relative risk are very similar. In these cases, the odds ratio can be roughly interpreted as relative risk (see homework).

Practice:

1-4) More Popcorn Problems
Reconsider the popcorn study of Investigation 1-1.
(a) Explain why it was appropriate to calculate the difference in proportions and the relative risk in the popcorn study but not for the Wynder and Graham study.
(b) Instead of conditioning on the level of exposure, calculate the conditional proportions of being in the high exposure group between those with airway obstruction and those without. Do these values give evidence of an association between these variables? Explain, being clear how you interpret these values.

Investigation 1-3: Lung Cancer and Smoking (Cont.)

Another landmark study investigating the link between lung cancer and smoking was conducted by Hammond and Horn (1958). Starting in 1952, they used 22,000 American Cancer Society volunteers as interviewers. Each interviewer was to ask 10 healthy white men between the ages of 50 and 69 to complete a questionnaire on smoking habits. Each year during the 44 month follow-up, the interviewer reported whether or not the man had died, and if so, how. They ended up tracking 187,783 men in nine states (California, Illinois, Iowa, Michigan, Minnesota, New Jersey, New York, Pennsylvania, and Wisconsin), over 44 months. Almost 188,000 were followed up by the volunteers through October 1955, during which time about 11,870 of the men had died, 448 from lung cancer. The following table (reproduced from the Hammond and Horn article) classifies the men as "having a history of regular cigarette smoking" or not and whether or not they died from lung cancer. Note that non-smokers are grouped with occasional smokers, including pipe- and cigar-only smokers.

	Regular smoker	Not regular smoker	Total
Lung cancer death	397	51	448
Alive or other cause of death	78557	108778	187335
	78954	108829	187783

(a) Identify the explanatory variable and the response variable. Also use Excel to make a segmented bar graph to examine the distribution of these data, putting the explanatory variable along the horizontal axis.

Graph:

Explanatory variable:

Response variable:

(b) Is this study best described as a case-control study, a cohort study, or were the subjects cross-classified? Explain.

Note: When we have categorized the observational units according to two variables, we use the same numerical summaries to analyze the *strength of the association* between the variables as to compare two groups on the same response variable.

(c) From the table above, calculate the difference in the proportion of "smokers" who died of lung cancer and the proportion of "non-smokers" who died of lung cancer. Does this difference seem large?

(d) From the table above, calculate the relative risk of lung cancer for regular smokers compared to the other group. Also calculate the odds ratio. What do these values tell you and do they seem large?

Relative risk:

Odds ratio:

(e) What are some advantages of this prospective cohort study compared to the retrospective case-control study in Investigation 1-2?

(f) Many critics still did not believe that this study established a causal link between smoking and lung cancer. What do you think some of their arguments were?

(g) Does the fact that a different study methodology led to a very similar comparison as the Wynder and Graham study strengthen the evidence of a causal link? Could there still be other explanations? If so, what are they?

Study Conclusions: Hammond and Horn concluded, like many other researchers of similar studies at the time, that the total death rate (not just for lung cancer) for cigarette smokers was approximately 70% higher than that for non-smokers. In all of the prospective cohort studies at this time, as the amount of cigarette consumption increased, so did total mortality rates. However, while the retrospective nature of the Wynder and Graham study is not an issue here, we must still have cautions in generalizing these results to a larger population (just how did these interviewers select the participants?) and in drawing cause and effect conclusions. Critics reasonably argued that other variables such as diet, exercise, and genetics could be responsible for both the smoking habits and the development of lung cancer. While there was still much (on-going) research to be done, and these studies did not claim to *prove* that cigarette smoking causes lung cancer, these landmark studies set the stage. They also led to many efforts in improving study design and in developing statistical tools (such as relative risk and odds ratios) to analyze the results.

EXCEL EXPLORATION

Return to the Excel spreadsheet you created in (a). It should look like:

	A	B	C
1		Regular smoker	Not regular smoker
2	Lung cancer death	397	51
3	Alive or other cause of death	78557	108778

(h) Recreate the segmented bar graph for these data.

(i) Set up formulas for computing the baseline risk of lung cancer death, the difference in the proportions of subjects who died from lung cancer between the regular smokers and the others, the relative risk of lung cancer death between the two groups and the odds ratio of lung cancer death for the regular smokers compared to those who did not smoke regularly. Your formulas should only make use of cells B2, C2, B3, and C3. For example, to calculate the proportion of "successes" in the "Regular smoker" group, you would type:

= B2/(B2+B3)

inside an empty cell. Note that the calculation cell starts with the equal sign. [*Hint*: You can greatly simplify the general formula for computing the odds ratio in this 2x2 table.]
Report the formulas you used:

Baseline risk:

Difference in proportions:

Relative risk:

Odds ratio:

Verify that the resulting values match your previous calculations of RR = 10.73 and OR = 10.78.

Now suppose the table had these hypothetical values:

	Regular smoker	Not regular smoker
Lung cancer death	39700	54226
Alive or other cause of death	39254	54603

(j) Replace the entries in B2, C2, B3, and C3 with these values. How does the segmented bar graph change? (Does the distribution look any different visually?)

How do the following values change?
Baseline risk:
Difference in proportions:
Relative Risk:
Odds Ratio:

Would you consider the risk of smoking as dramatic as above?

How do the values of the relative risk and the odds ratio compare to each other?

(k) Repeat (j) using the following hypothetical table:

	Regular smoker	Not regular smoker
Lung cancer death	63163	43532
Alive or other cause of death	15791	65297

      Baseline risk
      Difference in proportions
      Relative Risk
      Odds-Ratio

How do the values compare to those obtained from the table in (j) and to each other?

(l) Repeat (j) using the following hypothetical table:

	Regular smoker	Not regular smoker
Lung cancer death	31582	43532
Alive or other cause of death	47372	65297

What do you notice about the difference in proportions?

What do you notice about the value of the relative risk and the odds ratio?

Is lung cancer death as likely as being alive or dead from other causes?

Is lung cancer death among regular smokers as likely as among non-regular smokers?

Def: In a 2×2 table, when the conditional proportions are equal, we say that the variables are *independent* or not associated. In this case, the relative risk and the odds ratio must both be equal to 1. The further the relative risk and the odds ratio are from 1, the stronger the association/relationship/dependence is between the variables.

For the hypothetical data in (l), we would say that the smoking status was not related to whether or not the individuals died from lung cancer.

(m) Write a paragraph summarizing this exploration. What changed in each table and what effect did that have on the relative risk and odds ratio? When were the relative risk and odds ratio close in value? For what types of the baseline rate values will the relative risk and odds ratio be far apart? If the difference in proportions does not change, why might the relative risk and odds ratio be large or small?

Discussion: The difference between two proportions, relative risk, and odds ratios all provide a measure of the difference in the categorical response between two groups. One of the important observations you should make in this investigation is that focusing on the difference between the proportions of success is not always the best way to compare the two groups. Especially for characteristics with a low baseline rate, the relative risk and odds ratio provide a better comparison. Note that in the Hammond and Horn study, smoking regularly only reduced the proportion of lung cancer deaths by .0045, but this corresponds to more than 10 times as many subjects dying. The table in (j) had the same difference in proportions, but the relative risk and odds ratio indicate that this would correspond to only roughly 1% fewer deaths. This distinction is most noticeable when the proportions of success are close to 0 or 1 rather than close to 0.5.

One key observation from the table in (k) is that the proportions of "success" were quite high for both groups. In this situation, we tend to see a larger difference between the relative risk and the odds ratio values. But in (i) and in the Hammond and Horn study, the relative risk and the odds ratio values were rather similar. Recall from Investigation 1-2 that we cautioned that it is not appropriate to generalize the relative risk from a case-control study. This investigation shows that when the proportions of success are small in both groups, the odds ratio, which is always an appropriate summary, can be used to estimate the relative risk. With a cohort study such as the Hammond and Horn study, either the relative risk or odds ratio could be reported.

—

SECTION 1-2 SUMMARY

In this section you continued to practice calculating relative risk and odds ratio. It is important to keep in mind that while the odds ratio calculation is always meaningful, the relative risk calculation should not be used with case-control studies where the distribution of the response variable has been determined the study design. Oftentimes (when the success proportion is small), the calculated values are similar, but you must remember the slightly different interpretations. While these numerical summaries help you describe the strength of association in the experimental data, you must consider how the data were collected before attempting to draw and effect conclusions or to generalize results to a larger population. You will investigate these issues further in the next section and throughout the course.

Practice:

1-5) Minority Baseball Coaches

Journalist Sandy Tolan reported in the book *Hank and Me* that in 1999 there were fewer minorities coaching at third base than at first base. Tolan argued that since third base is the more challenging of the two positions and typically leads to more managing responsibilities, this discrepancy constitutes evidence of discrimination against minority coaches. Of the 60 base coaches in Major League Baseball that year (30 at first base and 30 at third), 21 were minorities. Of these 21, only 6 coached at third base.

(a) Fill in the following two-way table to represent these data:

	Minority coach	Non-minority coach	Total
First base coach			
Third base coach			
Total			

(b) Identify the observational units and the two variables represented here. Be sure to state the variables clearly as variables (as opposed to numerical summaries or conclusions or research questions). For each variable, specify whether it is quantitative or categorical.

(c) Which variable is being treated as the explanatory variable and which as the response variable?

(d) What proportion of the minorities coached at third base in 1999? What proportion of the non-minorities coached at third base? Construct a segmented bar graph to compare these groups.

(e) What proportion of coaches coached at third based in 1999? Explain why it is not appropriate to calculate the relative risk for these data.

(f) Calculate the odds ratio for coaching at third base in these two groups. Do these data appear to support the journalist's argument? Explain.

1-6) Violence Begets Violence

In a study of the hypothesis that victims of violence exhibit more violent behavior toward others, a researcher searched court records to find 908 individuals who had been victims of abuse as children (11 years or younger). She then found 667 individuals, with similar demographic characteristics, who had not been abused as children. She then searched through subsequent years of court records to determine how many of the people in each of these groups became involved in violent crimes (Data from Widom (1989) and Ramsey and Schafer (2002)).

		Abuse victim	Not abused
Involved in violent crime?	Yes	102	53
	No	806	614

(a) Identify the explanatory variable and the response variable for this study.

(b) Is this a case-control or a cohort study? Is this a prospective or a retrospective design? Explain.

(c) Based on your answer to (a), would it be appropriate to calculate the relative risk with these data? Explain.

(d) Calculate the odds ratio of involvement in violent crime to compare the abuse victims to the control group. Does this appear to support the researchers' conjecture? Explain.

1-7) Titantic Troubles

The following tables display the status of the passengers of the Titanic, their gender and whether they were a child or an adult, as determined by Dawson (1995). (The historical sources didn't completely agree with each other.)

Child	Male	Female
Survived	29	28
Died	35	17

Adult	Male	Female
Survived	338	316
Died	1329	109

(a) What was the odds ratio of survival for a child compared to an adult?

(b) What was the odds ratio of survival for a female compared to a male?

(c) Was the association between survival status and gender stronger for the children or for the adults? Justify your conclusion.

(d) Is it reasonable to conclude that adult females were about 11 times more likely to survive than adult males? Explain.

(e) Fill in the cells of the following (hypothetical) table so that the gender and survival status variables are independent.

	Male	Female	Total
Survived			58
Died			51
Total	64	45	109

(f) Calculate the odds ratio for this table that you created in (e). Does its value make sense? Explain.

SECTION 1-3: CONFOUNDING

A key consideration in this chapter is drawing appropriate conclusions from our data. When we see a big difference between two proportions or a large odds ratio, it can be very tempting to "jump to conclusions" and decide that the explanatory variable is *causing* the difference in the distributions of the response variable. More people die from heart attacks in November and December—does that mean that the holiday season is causing the higher number of heart attack deaths? In this section you will see some of the problems with drawing such conclusions.

Investigation 1-4: Near-Sightedness and Night Lights

Myopia, or near-sightedness, typically develops during the childhood years. Recent studies have explored whether there is an association between development of myopia and the use of night-lights with infants. Quinn, Shin, Maguire, and Stone (1999) examined the type of light children aged 2-16 were exposed to. Between January and June 1998, the parents of 479 children who were seen as outpatients in a university pediatric ophthalmology clinic completed a questionnaire (children who had already developed serious eye conditions were excluded). One of the questions asked was "Under which lighting condition did/does your child sleep at night?" before the age of 2 years. The parents chose between "room lighting," "a night light," and "darkness." Based on the child's most recent eye examination, they were separated into three groups: hyperopia (far-sightedness), emmetropia (normal refraction), and myopia. Of the 172 children who slept in darkness, 40 exhibited hyperopia, 114 normal refraction and 18 myopia. Of the 232 children with a night light, 39 exhibited hyperopia, 115 normal refraction, and 78 myopia. Of the 75 children who slept in a lit room, 12 exhibited hyperopia, 22 normal refraction, and 41 myopia.

(a) Identify the observational units and the two variables in this study. For each variable, specify whether it is quantitative or categorical.

Observational units:

Variables: type:

type:

(b) Which variable is being considered the explanatory variable and which is being considered the response variable?

(c) Is this study best described as a case-control study, a cohort study, or were the subjects cross-classified? Is the study retrospective or prospective? Explain.

(d) Construct a two-way table to summarize the data obtained by the researchers. Use the explanatory variable as the column variable.

(e) Construct (by hand or with Excel) a segmented bar graph to display the association between these variables. What does the bar graph reveal about whether myopia increases with higher levels of light exposure?

Note: If the bars have identical distributions, that is, if the breakdown among the eye condition groups is the same for each lighting condition, then the lighting condition and eye condition would be *independent*. (This is analogous to the definition of independence in a 2×2 table given in the previous section.)

(f) Determine the base rate of myopia with these children. Then determine the proportion of children with myopia in each lighting group. Finally, determine the proportion of children with hyperopia in each lighting group. Discuss what these results reveal about a relationship between the child's eye condition and the type of lighting.

Base rate of myopia:

Myopia with room light: with night light: with darkness:

Hyperopia with room light: with night light: with darkness:

Discussion:

(g) Based on these data, are you persuaded that the amount of light *caused* the higher rate of myopia in children? If not, suggest another explanation for the strong association observed by the researchers.

Discussion:

Since the proportion with each eye condition differs across the lighting groups, there appears to be an association between type of lighting and the children's eye condition. Still, the researchers caution that this study does not establish that the use of night lights *causes* the development of myopia in children. In particular, there are other factors that could determine why the parents chose the lighting condition that they did, such as socio-economic status of the family or the parents' own eye conditions. In turn, these factors are believed to be associated with myopia. For example, if the parents are near-sighted, they are more likely to use artificial lighting in their children's rooms and they are more likely to have near-sighted children.

Def: When another variable has an influence on the response, but its effects cannot be separated from those of the explanatory variable, we say the variables are *confounded*. When we classify subjects into different groups based on existing conditions (i.e., in an observational study), there is always the possibility that there are other differences between the groups apart from the explanatory variable that we are focusing on.

In the above study, it is possible that the parents' eye condition is confounded with the lighting condition. This would be true if near-sighted parents tend to use lighting more than parents with normal sight and if, in addition, near-sighted parents tend to have near-sighted children. In this situation, we would call the parents' eye condition a *confounding variable*. We would be much more convinced of a causal link due to the lighting condition if we knew that the children in the lighting conditions were otherwise similar.

A concern with all observational studies is the possibility of confounding variables. In fact, you will see in the next section that when we look at different subgroups of data, we may even see a reversal in the direction of the association due to a confounding variable. Thus, a large limitation with all observational studies is that we can never draw cause and effect conclusions. Evidence for causation can be strengthened when we see the association in the same direction in many different studies, when the association persists even after controlling for all known potential confounding variables, when a dose-response is shown, and when a scientific explanation is given. When all of these conditions are met, only then do scientists cautiously consider drawing a cause-and-effect relationship from observational studies.

Practice:

1-8) Holiday Season Heart Attacks
A study (referred to in the October 12, 1999 issue of *USA Today*) reported the number of deaths by heart attack in each month over a 12-year period in New York and Boston. December and January were found to have substantially higher numbers of deaths than the other months. Researchers conjectured that the stress and overindulgence associated with the holiday season might explain the higher numbers of deaths in those months compared to the rest of the year.
(a) Identify a confounding variable and alternative explanation for this result. Make sure you indicate how this confounding variable affects the two "treatment groups" differently.
(b) Explain why saying "people living in big cities have more stressful lives" does not identify a confounding variable here.
(c) Suggest a way of altering the study that would help isolate the effect of the holiday season.

1-9) Confounding Variables
For the following statements, identify which is being considered the explanatory variable and which the response variable. Then suggest a potential confounding variable that explains the observed association between the explanatory and response variables. Make sure you are clear how the suggested variable is related both to the response and to differences between the explanatory variable groups.
(a) Children with larger feet tend to have higher reading scores than children with smaller feet.
(b) Days with a higher number of ice cream sales also tend to have more drownings.
(c) Cities with higher teacher salaries also tend to have higher sales volumes in alcohol.
(d) People who eat apples regularly tend to have fewer cavities.
(e) Heart attack victims who have pets tend to live longer than those without pets.

Investigation 1-5: Graduate Admissions Discrimination

The University of California at Berkeley was charged with having discriminated against women in their graduate admissions process for the fall quarter of 1973. The two-way table below shows the number of men accepted and denied and the number of women accepted and denied for two of the university's graduate programs (Bickel and O'Connell, 1975).

	Men	Women
Accepted	533	113
Denied	665	336
Total	1198	449

(a) Calculate the proportion of men applicants who were accepted and the proportion of women applicants who were accepted. Is there evidence that men were accepted at a much higher rate than women, or do gender and admission decision appear to be independent? Explain.

Men:

Women:

(b) Does it look like Berkeley was guilty of gender discrimination? Explain.

(c) The table below identifies the number of acceptances and denials for both men and women applicants, broken down into the two graduate programs identified as A and F. (Notice that the column totals of the two programs match the counts in the two-way table above.) This is sometimes called a *three-way table*, because it displays results for three variables. List these three variables (like always, be sure to write them clearly as variables).

	Men		Women	
	Accepted	Denied	Accepted	Denied
Program A	511	314	89	19
Program F	22	351	24	317
Total	533	665	113	336

Variable 1:

Variable 2:

Variable 3:

(d) *Within each program*, calculate the proportion of men who were accepted and the proportion of women who were accepted. Did men have the higher rate of acceptance in both programs? Does this seem consistent with your results in (a)? Explain.

Program A   Program F
    Men: Men:

    Women: Women:

(e) There appears to be a "paradox" here—men have a higher acceptance rate overall, yet women have the higher acceptance rate within each program. Using the data provided in the table above, explain how the paradox happened in this study. *Hint*: What is the confounding variable here – what "lurking" variable reveals something about both gender and admission decisions that explains why the men had the higher acceptance rate overall?

Def: This unusual phenomenon is known as *Simpson's paradox*. It says that the direction of an association can reverse if you disaggregate the data according to a third variable. The explanation of Simpson's paradox always involves the fact that the previously "lurking" variable must be differentially related to both of the other variables.

In this example, men have a much higher acceptance rate than women overall, yet women have a higher acceptance rate than men in both programs. Here the explanation is that program A had much higher acceptance rates overall (than program F) and very few women applied to program A. In other words, the program variable is confounded with the explanatory variable of gender since women tended to apply to different programs than men, and the program variable also has an effect on the response variable of admission. In fact, this confounding variable led to a reversal in the direction of the association since the programs had such different acceptance rates.

(f) The program A acceptance rate for women was .824 and the program F acceptance rate for women was .070. Explain why it is not reasonable to average these values together to find the overall acceptance rate for women. Which value will the overall acceptance rate be closer to?

If we had only been given the acceptance rates and the application rates, we could still recombine this information to obtain the overall acceptance rates for each gender. The program acceptance rates would have to be "weighted" by the application rate for that gender:

Acceptance rate for women =
$$\left(\frac{\#women\,applied\,program\,A}{total\,\#\,female\,applicants}\right)\left(\frac{program\,A\,accept}{rate\,for\,women}\right) + \left(\frac{\#women\,applied\,program\,F}{total\,\#\,female\,applicants}\right)\left(\frac{program\,F\,accept}{rate\,for\,women}\right)$$

(f) Verify that this formula produces the overall acceptance rate from (a).

(g) Repeat (f) for the men.

(h) In the following table, the overall acceptance rate for women has not changed. Fill in this table so that the acceptance rate is the same for the two programs.

		Program A	Program F	Total
	Accepted			113
Women	Denied			336
	Total	108	341	449

(i) Does Simpson's Paradox occur for the table in (h)?

(j) In the following table, the overall acceptance rate for women has not changed. Fill in this table so that half the women apply to program A and half to Program F.

		Program A	Program F	Total
Women	Accepted			113
	Denied			336
	Total			449

(k) Does Simpson's Paradox occur for the table in (j)?

Discussion: Simpson's Paradox does not occur whenever the sample sizes in the two groups are similar or the acceptance rates in the two groups in similar. In either of these situations, the weighted average will simplify to be the average of the two acceptance rates. When we do have unequal sample sizes and success rates, then we should use a weighted average to combine the results across groups.

SECTION 1-3 SUMMARY

The principle of confounding reveals why one cannot draw cause and effect conclusions from observational studies: we are not able to eliminate the possibility of another variable that impacts the response variable in a way that is not distinguishable from the explanatory variable. A special case of this arises with Simpson's Paradox, and you worked through a few examples mathematically to see how the "paradox" can be explained.

Practice:

1-10) Kidney Surgeries

A recent study (Julious and Mullee, *British Medical Journal*, 1994) compared two types of procedures for removing kidney stones: open surgery and percutaneuous nephrolithotomy (PN), a "keyhole" surgery that removes the stone through the skin, designed to have much less disturbance than an open operation. In this study:

- For stones less than 2 cm large, 81 of 87 cases of open surgery were successful compared to 234 of 270 cases of PN.
- For stones of at least 2 cm large, 192 of 263 cases of open heart surgery were successful compared to 55 of 80 cases of PN.

(a) Is there evidence of Simpson's Paradox with this data? Include calculations to support your answer. If so, explain to someone not in a statistics class what is giving rise to the paradox in this study.

(b) Express the overall proportion of successful operations for the open surgeries as a weighted average of small stone and large stone surgeries.

1-11) On-time Arrivals

The following data report the number of flights that were "on-time" and "not on-time" for Continental Airlines and American Airlines in November, 2002 for all flights to Houston, Chicago, and Los Angeles.

	Houston		Chicago		Los Angeles	
	On time	Late	On time	Late	On time	Late
Continental	7318	1017	466	135	544	145
American	598	70	8330	1755	2707	566

(a) Calculate the on-time arrival rate for Continental for each city. Do the same for American. Which airline has a higher proportion of on-time arrivals in these cities?

(b) Calculate the overall arrival rate for each airline. Is Simpson's Paradox present with these data? Explain.

(c) Suggest an explanation for the cause of the paradox—what is the lurking variable here?

(d) Based on these data, which airline would you rather fly on? Explain.

SECTION 1-4: DESIGNING EXPERIMENTS

You saw in the previous section that confounding can present a very serious difficulty when trying to draw cause-and-effect conclusions. We can try to address confounding variables in the study design (e.g., seeing if heart attacks occur more often in November and December in the southern hemisphere as well), but it is not always possible to "control" for every other conceivable explanation. In this section you will learn how to design studies that address this issue and allow us to isolate the effects of our explanatory variable.

Investigation 1-6: Foreign Language and SAT scores

Some proponents of requiring students to study a foreign language contend that such study improves students' overall verbal ability. One way to investigate this conjecture is to examine the SAT-verbal scores of students who have studied a foreign language for at least two years and compare these scores to those of students who have never studied a foreign language.
(a) Identify the explanatory and response variables in such a study. Classify the variables as quantitative or categorical.

Explanatory:

Response:

(b) Suppose the study does find a large difference in the average verbal-SAT scores of these two groups. Explain why it would be problematic to conclude that this difference was entirely due to foreign language study. That is, suggest a confounding variable that might also explain why the students who studied a foreign language also had higher verbal-SAT scores. Clearly state how this variable distinguishes the explanatory variable groups and is linked to the response variable.

(c) Suggest how to design a better study to allow us to see whether foreign language study improves students' SAT-verbal scores.

Terminology Detour
In an *experiment*, the researchers actively impose the explanatory variable (also known as treatment variable) on the observational units. This active imposition differentiates experiments from *observational studies* where the researchers merely record information about the subjects without imposing the outcomes of the explanatory variable. In an experiment, you may want to refer to *experimental units* instead of observational units. When the experiment involves people or animals, the experimental units are then referred to as *subjects*.

A specific experimental condition applied to a subject is called a *treatment*. In a *comparative experiment* the researchers create at least two treatment groups or at least one treatment group and a control group (a group with no treatment or a "fake treatment" known as a *placebo*). For example, the researchers could take a group of students and assign half to study a foreign language and the other half not to study a foreign language. Still, the researchers must also take steps to ensure that the created groups are as similar as possible with regard to other variables.

(d) If you were to conduct an experiment to investigate whether foreign language study causes an increase in verbal SAT scores, what would you like to be true about the treatment groups?

One way to achieve this goal of creating treatment groups that are as similar as possible is through *randomization*, using a random mechanism to decide which subjects are assigned to which treatments.

A schematic to represent this design could be:

Investigation 1-7: Have a Nice Trip

An area of research in biomechanics and gerontology concerns falls and fall-related injuries, especially for elderly people. Recent studies have focused on how individuals respond to large postural disturbances (e.g., tripping, induced slips). One question is whether subjects can be instructed to improve their recovery from such perturbations. Suppose researchers want to compare two such recovery strategies, lowering (making the next step shorter, but in normal step time) and elevating (using a longer or normal step length with normal step time). They will induce a trip while the subject is walking (but harnessed for safety) using a concealed, mechanical obstacle. Subjects will be trained on the recovery strategy, and asked to apply it after they feel themselves tripping.

Suppose the following 12 subjects have agreed to participate in such a study:

> Females: Audrey, Mary, Barbie, Anna
> Males: Matt, Peter, Shawn, Brad, Michael, Kyle, Russ, Patrick

(a) Suppose the four women volunteer to try the lowering strategy and the eight men volunteer to try the elevating strategy. Would this be a problem? Explain.

(b) How could we fairly decide which subjects are instructed on the lowering strategy and which on the elevating strategy? When we have created the two groups, what would we like to be similar about the heights of the subjects in the two groups? About the gender breakdown of the two groups? Why is this important?

A key concern here is that the gender breakdown is similar for the two groups. Notice, we are not trying to get an equal number of men and women in Group 1, but we would like the proportion of men in Group 1 to be equal to the proportion of men in Group 2 (so gender and group membership are independent). While we could force the gender breakdown to be the same in both groups, there are likely to be other variables that may also be related to the response variable. We would like a method of assigning subjects to groups has a high likelihood of creating similar groups on all variables, whether we can identify or even measure them at the start of the study.

In this investigation, you will randomly assign the subjects to groups to see how often similar groups are created. We will focus on a variable like gender in order to help you see how the randomization process works. You will investigate the distribution of the difference in proportions of males between the two groups as you repeat the randomization process.

(c) Take 12 index cards and write each of the 12 names on a card. Also keep track of the gender of each subject. Shuffle the cards well and then deal out 6 of them. These 6 will be instructed to use the lowering recovery strategy. The other six will be instructed to use the elevating strategy.

How many men did you end up with in the lowering recovery strategy group? How many women? What proportion of the lowering recovery strategy group is male? What proportion of the elevating group is male? Record your results in the table below and then repeat your randomization a second time.

	Lowering Strategy			Elevating Strategy		
	# men	# women	Proportion that are male	# men	# women	Proportion that are male
Trial 1						
Trial 2						

(d) Calculate the difference in the proportion of men in the two groups (always subtracting in the same direction, "lowering"-"elevating") for each trial. Pool your results with your classmates to create a *dotplot* that represents the distribution of these differences. Place a dot to correspond to each individual repetition, stacking the dots for multiple observations at a particular value.

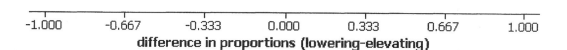

(e) Did randomization equally distribute the men and women between these two groups? If not, is the gender variable reasonably balanced between the two groups? How do your results compare to the rest of the class?

To see how well randomization "works" we must examine the long-run behavior: what is true about the equality or balance between the groups if we repeat this process many, many times? To expedite this exploration, we will use a java applet to repeat the randomization process for us.

(f) Open the "Randomization of Subjects" applet and click the Randomize button.

What proportion of subjects assigned to Group 1 are men? Of Group 2? What is the difference in these two proportions?

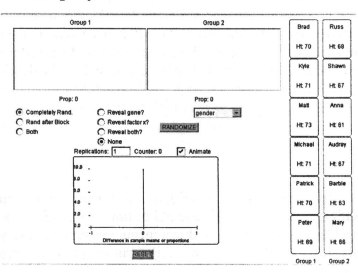

Randomizing Subjects

You will notice that the difference in proportions is shown in the bottom graph, called a *dotplot*. Each dot represents the difference in the proportions of men between the two groups for one randomization. In other words, the observational unit in that graph is a repetition of the random assignment, and the variable is the difference in proportions of men between the two groups.

(g) Click the Randomize button again. Was the difference in proportions the same this time?

(h) Change the number of replications from 1 to 48, click off the Animate option, and click the Randomize button. The dotplot will display the difference between the two proportions for each of the 50 repetitions. Do you see any pattern in the differences? Describe this pattern.

(i) Change the number of repetitions from 48 to 150 and click the Randomize button. The dotplot now shows the differences for 200 repetitions of this randomization process. Where are these values centered? What was the largest difference ever obtained ? (If you click on a dot in the dotplot, the applet will redisplay the randomization that led to that difference.) What was the smallest difference ever obtained?

Center: Largest: Smallest:

(j) Does randomization *always* equally distribute/balance the men and women between the two groups? Is there a tendency for there to be a similar proportion of men in the two groups? Explain.

(k) Prior research has also shown that the likelihood of falling is related to variables such as walking speed, stride rate, and height, so we would like randomization to distribute these variables equally between the groups as well. In the applet, use the pull-down menu to switch from the gender variable to the height variable. The dotplot now displays the difference in average height between Group 1 and Group 2 for these 200 repetitions. In the long-run, does randomization tend to equally distribute the height variable between the two groups? Explain.

(l) Suppose there is a "balance gene" that is related to people's ability to recover from a trip. We didn't know about this gene ahead of time, but if you click the "reveal gene" button and then select "gene" from the pull down menu, the applet shows you how the proportions with the gene differ in the two groups. Does this variable tend to equalize between the two groups in the long run? Explain.

(m) Suppose there were other "*x*-variables" that we could not measure such as stride rate or walking speed. Select the "reveal both" button and use pull-down menu to display the results for the *x*- variable. Does randomization generally succeed in equalizing this variable between the two groups? Explain.

Discussion: The primary goal of *randomization* is to create groups that have similar characteristics/distributions prior to imposing the explanatory variable. Thus, randomization equalizes the effects of any potential confounding variables between the groups, creating groups that overall only differ on the explanatory variable imposed. Note that this "balancing out" applies equally well to variables that can be observed (such as gender and height) and variables that may not have been recorded (such as age or walking speed) or that can not be observed (such as a hidden gene). While we could have forced variables like gender and height to be equally distributed between the two groups, the virtue of randomization is that it also balances out variables that we might not have thought of ahead of time or and variables that we might not be able to see or control.

Thus, if we observe a "significant" difference in the response variable between the two groups at the end of the study, we feel comfortable attributing this difference to the explanatory variable (e.g., recovery strategy employed) as that should have been the only difference between the groups. This allows us to draw a cause-and-effect conclusion between the explanatory and the response variable. We are not able to draw such conclusions from observational studies, because they always have the potential of confounding variables.

Terminology Detour

Def: A *randomized comparative experiment* imposes two or more treatments on the subjects and through randomization gives each subject an equal chance of being in any of the treatment groups. This process tends to create groups that are similar on all pre-treatment variables so that if we observe a significant difference in the response variable between the groups at the end of the study, we can attribute the difference to the explanatory variable. These types of studies, brought to the forefront of scientific attention by R.A. Fisher (1935), are considered the gold standard for determining cause and effect relationships between variables.

Keep in mind that while this is the ideal, it is not always feasible or ethical to randomly assign subjects to treatment groups.

(n) For example, do you think we could conduct a randomized comparative experiment for the myopia study? For the foreign language study? For the smoking study? Explain.

Statisticians will often employ other strategies for ensuring comparability between groups.

(o) For example, suggest a problem with telling those that they have been assigned to take a foreign language that they are suspected to perform better and telling those that have been assigned not to take a foreign language that they are expected to perform more poorly.

Def: If possible, you should *blind* the subjects to which treatment group they are in and to the researchers' expectations. This is often done through a *placebo treatment*. This fake treatment should be designed to appear exactly as the treatment of interest but to have no substantial effects, e.g., comparing the effects of aspirin to an empty pill of the same size, color, and texture. A *double blind experiment* blinds both the participants and the individuals who will be evaluating the response variable for the subjects.

Discussion: For example, the evaluators who grade the verbal ability exams should not know which students took the foreign language. A slightly artificial example, but you should certainly agree that if the researchers want to determine whether subjects "feel better" or "look happier" on a certain treatment, we need to prevent any hidden 'biases' on the part of the evaluator in making these subjective measurements. Ideally subjects will not even know they are in an experiment or even being observed as this knowledge can also affect their behavior. This last statement is often referred to as the "Hawthorne Effect," named after the Hawthorne Plant of Chicago's Western Electric Company where it was first observed in the late 1920s. In the next section you will learn about another technique for trying to ensure balance between the groups.

(p) Could a placebo treatment be designed for the foreign language study? For the smoking study?

Practice:

1-12) Positive Thinking
In a study published in the July 2003 issue of the journal *Psychosomatic Medicine*, researchers reported that people who tend to think positive thoughts catch a cold less often than those who tend to think negative thoughts. The scientists recruited over 300 initially healthy volunteers, and they first interviewed them over two weeks to gauge their emotional state, eventually assigning them a numerical score for positive emotions and a numerical score for negative emotions. Then the researchers injected rhinovirus, the germ that causes colds, into each subject's nose. The subjects were then monitored for the development of cold-like symptoms. Subjects scoring in the bottom third for positive emotions were three times more likely to catch a cold than those scoring in the top third.
(a) Identify the explanatory and response variables in this study. Classify each as categorical or quantitative.
(b) Is this an observational study or an experiment? Explain. [*Hint*: Ask yourself whether the explanatory variable was assigned by the researchers or not.]

1-13) Have a Nice Trip (cont.)
 For the 12 subjects in the tripping experiment,
(a) One could suggest tossing a coin 12 times and if the coin lands heads, sending the person to Group A and if the coin lands tails, sending the person to Group B. Discuss a minor weakness in this randomization method. Do you think this weakness could lead to confounding? Explain.
(b) Explain how you could use a coin to carry out the randomization to achieve an equal number of subjects in the two groups. Discuss a weakness in this randomization method. Do you think this weakness could lead to confounding? Explain.

1-14) Designing Studies
For each of the following research questions, describe how you would design an observational study to address the question and how you would design an experiment. In each case, identify which study you feel is more appropriate. In particular, are there any ethical or logistical issues that might prevent you from carrying out the experiment?
(a) The effects of second-hand smoke on the health of children.
(b) Whether people tend to spend more money in stores located next to food outlets with pleasing smells.
(c) Whether cell phone use increases the rate of automobile accidents.
(d) Whether people respond differently to two phrasings of the same survey question.

1-15) Designing Studies
For the studies in Practice 1-14 that you decided could be conducted as an experiment, explain how you could make the subjects blind to which treatment group they are assigned to.

Investigation 1-8: Have a Nice Trip, cont. (Optional)

(a) Can you suggest a modification to the "nice trip" study design and how we randomly assign subjects to the treatment conditions that would ensure balance of important variables like gender?

Def: A *generalized randomized block design* first splits the subjects into groups according to a *blocking variable* and then carries out the randomization to treatment groups separately within each block. We consider blocking on a variable if we strongly suspect that by doing so we will create more homogenous treatment groups.

Here, we would first separate the men and women since we expect the men are more similar to each other than to the women. Then randomly assign the men to different treatment groups and then separately randomize the women to the treatment groups. This allows us to compare the effect of the tripping strategy among the men, without the "noise" of differing balancing abilities between men and women.

No 2

(b) Separate the names on the 12 index cards into their gender groups. Randomly shuffle the men's cards and deal four of them to Group 1. Then randomly shuffle the women's cards and deal two of them to Group 1. How does the proportion of men compare between the two treatment groups?

(c) In the randomization applet, select the "Randomize after Blocking" button. Then have the applet perform 200 replications and examine the distribution of the differences in the proportion of men in each group. Explain why this dotplot is so boring.

(d) Now examine the dotplot of the difference in average height between the two groups. How does it compare to the distribution when we did not block on gender? (If you click the "both" button, you will see the distribution without blocking (in grey) and the distribution after blocking by gender (in black) at the same time.)

(e) Explain why it makes sense that the differences in the average heights are closer to zero when we block on gender.

(f) Reveal the gene and x-variables and look at the gene dotplot and the x-variable dotplot (for both randomization methods). How do these distributions compare with and without blocking? Did blocking create groups that are more similar with respect to these variables? What does this tell you about the difference in the proportion with the hidden gene between men and women and in the average of the x-variable between men and women? Did blocking create groups that tended to be less similar? Explain.

If there is a variable that you believe is related to the response variable and can be easily determined, you can take that variable into account prior to imposing the treatment. By blocking on this variable, you have ensured that the treatment groups are more homogeneous than if you didn't block on that variable. For example, blocking on gender also made the distribution of heights more similar between the two groups. Gender isn't strongly related to the gene variable or the x-variable so blocking on gender did not really change those randomization distributions, though it did not hurt either. In this latter case, blocking may not noticeably improve the efficiency of the study, although it won't decrease the efficiency either.

Practice:

1-16) Restaurant Spending and Music
A British study examined whether the type of background music playing in a restaurant affected the amount of money that diners spent on their meals. The researchers asked a restaurant to alternate silence, popular music, and classical music on successive nights over eighteen days. They found that patrons spent more when classical music was playing.
(a) Identify the explanatory and response variables in this study. Classify each as categorical or quantitative.
(b) Is this an observational study or an experiment? Explain.
(c) The experimenters did not use randomization in this study. Explain why this presents a problem for their conclusion by identifying a potentially confounding variable that may not have been balanced out among the treatment groups.
(d) Suggest a potential blocking factor for this study.

1-17) Foreign Language Study (cont)
 For the foreign language study, suggest a variable that the researchers might choose to block on and explain why this might be useful.

SECTION 1-5: ASSESSING STATISTICAL SIGNIFICANCE

You saw in the previous section that a key advantage of using randomization to create treatment groups is to feel comfortable that you have created similar groups prior to imposing your treatment, allowing you to draw cause-and-effect conclusions afterwards. Still, we don't expect all variables to be exactly equalized between the two groups, and we may observe some differences in the response between the two groups even if there is no effect from the treatment. The second, and equally important, key advantage to using randomization is that we will be able to predict how different results in a response variable will be between experimental groups just due to the chance nature of the random assignment. This allows us to draw conclusions about whether an observed difference is too big to have reasonably occurred just due to chance. We will use two primary strategies for addressing this question: 1) simulation, and 2) a mathematical probability model.

Investigation 1-9: Friendly Observers
In a study published in the *Journal of Personality and Social Psychology* (Butler and Baumeister, 1998), researchers investigated a conjecture that having an observer with a vested interest would decrease subjects' performance on a skill-based task. Subjects were given time to practice playing a video game that required them to navigate an obstacle course as quickly as possible. They were then told to play the game one final time with an observer present. Subjects were randomly assigned to one of two groups. One group (A) was told that the participant and observer would each win $3 if the participant beat a certain threshold time, and the other group (B) was told only that the participant would win the prize if the threshold were beaten. The threshold was chosen to be a time that they beat in 30% of their practice turns. The following results are very similar to those found in the experiment: 3 of the 12 subjects in Group A beat the threshold, while 8 of 12 subjects in Group B achieved success.

	A: observer shares prize, has vested interest	B: no sharing of prize, no vested interest	Total
Beat threshold ("success")	3	8	11
Did not beat threshold ("failure")	9	4	13
Total	12	12	24

(a) Explain why this is an experiment and not an observational study. Was the experiment *double blind*? Explain.

(b) What are the experimental units in this study, and what are the variables? Also identify the variables as quantitative or categorical. Finally, specify which is the explanatory and which is the response variable.

(c) Calculate the proportion of successes (beating the threshold) for each group and compute the odds ratio of beating the threshold between the two groups.

(d) Produce a segmented bar graph to display the experimental results, placing the explanatory variable along the horizontal axis.

(e) Calculate the difference in the proportion of successes between these two groups. Do the proportions of success observed by the researchers differ in the direction they conjectured beforehand?

(f) Even if the observer's interest had absolutely no effect on subjects' performance, is it *possible* to have gotten a difference like this just due to chance variation? Explain.

From these data, we see that some subjects beat the threshold and some did not. As the researchers conjectured, a higher proportion of Group B subjects beat the threshold. However, recall from the last section that when we randomly assign subjects to groups, there is always the possibility that some characteristics of the groups will not match exactly, just "by chance." But here's where the second benefit of randomness comes in: it allows us to anticipate how large a difference we can expect to see in the two groups just by chance. Then, if we decide that the difference we observed between these two groups is too big to happen often by chance, we will say the difference between the two groups is "statistically significant."

To decide whether the difference is statistically significant, we will initially assume that there is no effect of the observer with the vested interest. In other words, we start by assuming that the subjects who beat the threshold and those who did not were going to do so regardless of which group they were assigned to, that the group assignment and their performance are independent.

Since this study saw 11 successes and 13 failures overall, then through the randomization process we would have expected roughly 5 or 6 successes in Group A and 5 or 6 successes in Group B. But we know that sometimes the randomization would lead to a larger difference between the groups. The key question is then: how likely is it for the randomization process to result in as few as 3 of the successes in Group A? (Note, once we know how many successes are in Group A, we know the results for the rest of the table since the 11 successes and 13 failures overall are fixed. So looking at the number of successes in Group A also tells us about the number of successes in Group B and therefore the difference in the proportion of successes between the two groups.)

One way to analyze this question is to *simulate* the randomization process carried out by the researchers. We will randomly assign 12 of the subjects to Group A and note how often we obtain a result at least as extreme as in the actual experiment. Since the researchers conjectured that there would be fewer successes in Group A, we will consider results of 3 or less as support for their conjecture. Repeating the simulation a large number of times will give us a sense for how unusual the researchers' result (or ones even more extreme in the direction of their conjecture) would be to occur by chance alone.

(g) Mark 11 cards as "success" and 13 as "failure," shuffle them well, and randomly deal out 12 to represent the subjects assigned to Group A. How many of these 12 are successes? Is this result as extreme as in the actual study?

(h) Repeat this randomization process a total of five times, recording your results in the table:

Repetition #	1	2	3	4	5
"successes" assigned to Group A					
as extreme as actual sample? (Y or N)					

(i) On the axis below, create a dotplot of the results for your whole class.

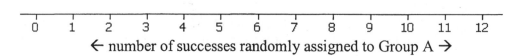

$\leftarrow$ number of successes randomly assigned to Group A $\rightarrow$

(j) How many repetitions were performed by the class as a whole? How many of them gave a result as extreme as the actual sample (3 or fewer successes in group A)? What proportion of the repetitions is this?

     Number of repetitions:
     Number as extreme:
     Proportion as extreme:

Practice:

1-18) More on Friendly Observers
Reconsider your simulation analyses of the psychology experiment.
(a) Suppose that the experimenters had found only 2 successes in Group A and 9 in Group B (for the same total of 11 successes). Use your applet tally results to approximate the p-value of this result. Is this p-value smaller or larger than before? Explain how your interpretation of the experimental results, with regard to the strength of evidence that the observer's vested interest causes a decrease in the subjects' performance, differs in this situation.
(b) Repeat (a) supposing that the experimenters had found 4 successes in Group A and 7 in Group B.

1-19) Relieving Back Pain
A study published in the journal *Neurology* (May 22, 2001) examined whether the drug botulinum toxin A is helpful for reducing pain among patients who suffer from chronic low back pain. The thirty-one subjects who participated in the study were randomly assigned to one of two treatment groups: 16 received a placebo of normal saline and the other 15 received the drug itself. The subjects' pain levels were evaluated at the beginning of the study and again after eight weeks. The researchers found that 2 of the 16 subjects that received the saline experienced a substantial reduction in pain, compared to 9 of the 15 subjects who received the actual drug.
(a) Is this an experiment or an observational study? Explain.
(b) Explain the importance of using the "placebo treatment" of saline in this study.
(c) Create the two-way table for summarizing these data, putting the explanatory variable as the columns and the response as rows.
(d) Calculate the conditional proportions of pain reduction in the two groups. Display the results in a segmented bar graph. Comment on what this preliminary analysis reveals.
(e) If there was no association between the treatment and the back pain relief, about how many of the 11 "successes" would you expect to see in each group? Did the researchers observe more successes in the saline group than expected (if the drug had no effect) or fewer successes than expected? Is this in the direction conjectured by the researchers?
(f) Open the Friendly Observers applet but follow the link at the bottom to be able to change the table entries. Use the applet to simulate 1000 random assignments of these subjects to the experimental groups. Use the placebo group as Group A. Assume that botulinum is not effective and so the 11 who experienced a reduction in pain would have done so regardless of which group they were in. Draw a sketch of the resulting randomization distribution, and report the approximate p-value. [Recall: The p-value is how often the randomization process alone produces a result as extreme as what the researchers obtained or more extreme in the direction conjectured by the researchers, if there really was no difference between the groups.]
(g) Summarize your conclusion about the effectiveness of botulinum. Be sure to refer to your simulation results, and also explain the reasoning process that leads to your conclusion.

BRIEF INTRODUCTION TO MINITAB

The Minitab[R] statistical package will provide extensive support to the analyses and investigations that you will be performing in this class. In these materials, we will be assuming that you have access to version 14 of Minitab. Much of the functionality is the same, but there are places where the command structure is slightly different in older versions of Minitab.

When you first open Minitab, you will notice two windows, a *session window* (top window) and a *data window* (bottom window). The data window looks like a spreadsheet but has very limited capabilities. Typically each column represents a variable and each row represents an observational unit. You will pass commands to Minitab, both through menus and by typing commands into the session window. In the future, you will have the choice of which way is more convenient for you. When you are done with this activity, you might want to look back through it and compile a list of all the features of Minitab that you used.

Saving Minitab output is done primarily through *Minitab worksheets* and *Minitab projects*. A Minitab worksheet (.mtw) contains just the data window. A Minitab project (.mpj) contains the data window, the session window, and any graphs that you have created. Initially data files will be given to you as worksheets but you may prefer to save your work as a project file. You should also get into the habit of transferring the relevant output from Minitab into a word processing program in order to produce well-written reports.

You will often have the choice of using the Minitab menus to access commands or to type commands directly into the Session window. In order to type in the Session window, you first need to *enable commands*. To do this, click in the Session window and then choose Editor > Enable Commands. You should now see the Minitab prompt MTB> in the Session window. You will need to do this each time you start Minitab. [To turn the MTB prompt feature on permanently, choose Tools > Options, select Session Window, then Submitting Commands. Click the button next to Enable and click OK. Alternatively, you can use point-and-click menus to access all commands.

Your first use of Minitab in the next section will be to run a simulation. Minitab *macros* provide a mechanism for repeatedly running the same Minitab commands a large number of times. We will initially tell you what commands to use, but you should always make sure you understand the functionality of the commands, even typing a few of them in directly in the session window to see what they do. Minitab macros are created with a text editor, like Notepad, and are most conveniently accessed if they have the extension ".mtb". You can ensure that the text editor maintains this extension by putting quotes around the file name when you save the file. Then you need to save the file in a memorable location so that you can later access it from within Minitab.

In Chapter 2 you will learn much more about Minitab data analysis tools. Minitab will be used extensively throughout this book, so we encourage you to get comfortable with it early and to keep a record for yourself of what the various Minitab commands do, for later reference.

Investigation 1-11: Selecting Senators

In the previous investigation there were only 15 outcomes in our sample space and it was feasible to list each outcome. Suppose senators were suspicious about the gender subcommittee that had been selected. The 2004 U.S. Senate consists of 86 men and 14 women and they are randomly assigned to a subcommittee of 5 members and a second group of 95 members. Let the random variable X represent the number of women in the subcommittee.

(a) What are the possible values for X, the number of women randomly assigned to the subcommittee?

(b) Which of these possible values do you think is the most likely outcome? Explain.

(c) How many possible subcommittees of 5 can be obtained from this group of 100 senators?

[Note: You can use Excel to carry out this calculation for you by entering "=combin(100,5)" in an empty cell.]

Clearly we don't want to have to list out all 75,287,520 possible outcomes and count the number of women in each one. Suppose we are interesting in the probability that exactly one woman is selected, $P(X=1)$. What we really need to know is how many of the subcommittees consist of exactly 1 woman.

(d) How many ways are there to select one woman from the 14 in the 2004 senate? [Use a combination.]

(e) Explain why the probability of obtaining exactly one woman on the committee is not simply your answer to (d) divided by your answer to (c).

Your analysis so far has overlooked the fact that if exactly one woman is chosen for the committee, then exactly four men must be chosen. If Hillary Clinton was chosen for the subcommittee, you still need to consider *all* possible committees that consist of her and 4 different men.

(f) Use a combination to determine the number of ways to select 4 men from the 86 male senators.

(g) To determine the total number of ways to select the one woman and the 4 men, multiply your answers to (d) and (f).

(h) Divide your answer to (g) by your answer to (c) to determine the probability of obtaining one woman and four men on the subcommittee.

(i) Repeat this analysis to determine the probability of obtaining two women and three men on the subcommittee. [*Hint*: One component of this calculation will not change.]

(j) Rather than repeat this analysis for all possible values of the number of women selected, let's try to derive a general expression for any value of x. Follow your analysis above to state a general expression for the probability of obtaining x women (and therefore 5-x) men on the subcommittee.

$P(X = x) =$

(k) How would the above expression change if there were r women in the 2004 senate?

(l) How would the above expression change if there were N members in the 2004 senate?

You have derived the formula for calculating *hypergeometric probabilities*. If you have a group of N objects where each object can be classified as a success or a failure and you plan to randomly select n of the objects and count the number of successes, the probability of obtaining x successes is equal to:

$$P(X=x \text{ successes}) = \frac{\text{number of ways to select } x \text{ successes and } n-x \text{ failures}}{\text{number of ways to select } n \text{ objects}} = \frac{\binom{M}{x}\binom{N-M}{n-x}}{\binom{N}{n}}$$

The *parameters* values of this hypergeometric probability distribution are N, M, and n. Specifying these values indicates exactly which hypergeometric distribution you are working with. It can be shown that each of these probabilities is between 0 and 1, inclusive, and that the sum of these probabilities over all possible values of x is one.

Note: In many textbooks and computer programs, N is referred to as the population size and n is referred to as the sample size.

It can also be shown that the theoretical expected value of this random variable is equal to the sample size times the proportion of successes in the population: $E(X) = n(M/N)$

(k) Use Minitab to perform these calculations:

- Put the values 0, 1, 2, 3, 4, and 5 into C1 and name this column "x"
- Select Calc > Probability Distribution > Hypergeometric
  - Click the "Probability" option
  - Enter the appropriate parameter values for the senate subcommittee example
  - Click on the "input column" and enter c1 for that column
  - Specify "prob" as the optional storage column (Minitab will use and name C2 for you).
- Make a graph of this probability distribution:
  MTB> plot c2*c1;
(puts probabilities on vertical, *x* values along horizontal and drops down vertical lines).

```
SUBC> project.
```

Report the probability distribution (also known as *probability mass function* or *pmf*) in the following table.

# women	0	1	2	3	4	5
Probability						

Verify that the probabilities reported by Minitab match the two you calculated above.

(l) What do you notice about the sum of these probabilities (MTB> sum c2)?

(m) Calculate the expected value of this random variable, $E(X) = \Sigma x P(X=x)$:

```
MTB> let c3=c1*c2
MTB> sum c3
```

Confirm that this corresponds to the expected value formula given for a hypergeometric random variable.

    Minitab:

    n(M/N):

(n) What is the most probable outcome? How was your guess in (b)? Is the most probable outcome equal to the expected value?

    Most probable:

    Guess in (b):

    Compared to expected value:

(o) What is the probability that women outnumber men on the subcommittee? [*Hint*: To find the probability of this *event*, add the probabilities of the outcomes comprising this event.]

(p) Do you expect that the probability of women outnumbering men would be larger, smaller, or the same, if a committee of size three (rather than five) were chosen? Explain.

(q) Use the hypergeometric probability distribution to calculate the probability in (p) and comment on whether your prediction is confirmed. [*Hint*: Feel free to use Minitab. If there are just a few *x* values that you are interested in, you can enter them individually in the "input constant" box instead of creating the entire probability distribution. So your first step is to figure out which values of *x* correspond to the event of interest here.]

    Probability women outnumber men with *n*=3:

    Comparison to prediction in (p):

No2

Practice:

1-20) Random Babies
Suppose that on one night at a certain hospital, four mothers give birth to baby boys. As a very sick joke, the hospital staff decides to return babies to their mothers completely at random.
(a) Let X represent the number of mothers that receive their correct baby. Explain why X does *not* follow a hypergeometric distribution.
(b) List all of the outcomes in the sample space. Use the notation "abcd" to mean that mother 1 gets baby a, mother 2 gets baby b, mother 3 gets baby c, and mother 4 gets baby d. So, two possible outcomes are 1234 and 1243. List the possible outcomes systematically so that you do not double-count or miss any.
(c) For each outcome in the sample space, determine how many mothers get the right baby with that outcome. Note: the outcome 1234 has all four mothers receiving the correct baby, 1243 has only the first two mothers receiving the correct baby.
(d) Use your answer to (c) to determine the probability of 0 correct matches. Then repeat for all possible values for the number of correct matches and report the probability distribution for X, the number of mothers who receive the correct baby.
(e) Open the "Random Babies" applet. Click the Randomize button to assign babies to mothers. How many mothers got the right baby?
(f) Repeat this randomization a second time. Did you get the same result?
(g) Change the number of trials to 98 and click off the Animation box. In the Cumulative Results box, what is the relative frequency of 0 correct matches? How does this compare to your answer to (d)?
(h) Click the bar in the graph corresponding to 0 matches. This displays the relative frequency over time. Does the relative frequency appear to be converging yet? If not, continue to click the Randomization bottom until you believe the relative frequency is close to its limit. How many repetitions did you use and what is that limiting value? How does this limiting value compare to the probability you calculated in (d)?
(i) Click the button to toggle back to the Histogram and then click the button to toggle to the Average. This displays the average number of mothers receiving the correct baby over time. Does this average appear to be converging yet? If not, continue to click the Randomization bottom until you believe the average is close to its limit. How many repetitions did you use and what is that limiting value?
(j) Use the probability distribution in (d) to determine the expected value of the random variable X. How does this expected value compare to the limiting value you found in (i)?

1-21) Randomizing Subjects
Recall that in Investigation 1-7 you randomly assigned twelve subjects (four women and eight men) to two treatment groups of six subjects each. In that activity you used index cards to perform the random assignment and then used an applet to repeat the randomization a large number of times. The point was to see how well randomization balances out various factors (such as gender and height) between the two treatment groups. Now you can use hypergeometric probabilities to analyze what would happen in the long run.
(a) How many different ways are there to make the random assignment of 12 people into two groups of 6 each? [*Hint*: This is the same as the number of ways of choosing 6 people from a group of 12 people.]

(b) Let the random variable X represent the number of women in Group A. What are the possible values for X?

(c) Calculate the hypergeometric probability of obtaining 2 women and 4 men in Group A, $P(X=2)$.

(d) Let Y represent the number of men in Group A. What is the hypergeometric probability of obtaining 4 men and 2 women in Group A, $P(Y=4)$?

(e) Returning to the number of women in Group A, calculate the hypergeometric probability for each possible value of X, the number of women in Group A, specified in (b). [*Hint*: You may want to use Minitab to carry out the calculations.]

(f) Make a graph of this probability distribution and calculate the expected number of women in Group A, $E(X)$.

(g) Based on your probability distribution, is it more likely that the genders will be equally divided between the two groups or that they will not? Explain.

(h) Determine the probability that randomization puts all of the women in one group? How surprising is this event? [*Hint*: Either Group A or Group B could be the group with all women.]

(i) How many men would you need to see in Group A before you would start to become skeptical that the outcome arose by chance alone? [*Hint*: For what value of y is $P(Y \geq y) \leq .05$?]

SECTION 1-7: FISHER'S EXACT TEST

Recall that we started analyzing the psychology experiment by considering how to decide if the difference in experimental outcomes between two groups was large enough to be deemed "significant." We saw that the randomization used to assign subjects to treatments enables us to determine the significance of the result. We used simulation to replicate the randomization process over and over, focusing on how often randomization alone produced a difference as extreme as the one actually obtained.

Simulation is a powerful tool, but it only produces approximate answers. We can do better than that with a formal mathematical analysis. Toward that end, in the last section we studied some basic ideas of probability and learned some counting rules that facilitate the calculation of probabilities. Now we are ready to return to the statistical issue and apply a probability analysis to it.

Investigation 1-12: More Friendly Observers

Recall that the psychology experiment involved 24 subjects, 12 of whom were randomly assigned to each of two groups.

(a) How many different ways are there to do this random assignment? [*Hint*: Think of this as randomly choosing 12 subjects from the 24 for assignment to Group A.] Would you like to list all of these outcomes?

Recall that the experiment resulted in 11 "successes" and 13 "failures," with three of those successes in Group A.

(b) Let X represent the number of successes randomly assigned to Group A. Assuming that there were no difference between the groups and so randomization alone accounted for how many successes were assigned to each group, determine the probability of obtaining 3 successes and 9 failures in Group A, P(X=3). [*Hint*: Use a hypergeometric probability.]

(c) Also determine the probability of getting 2 successes in Group A, the probability of getting 1 success in Group A, and the probability of getting 0 successes in Group A.

P(X=2):

P(X=1):

P(X=0):

Recall that the *p-value* is the probability of getting a result at least as extreme as the observed result, in the direction conjectured by the researchers. Here, the researchers believed there would be fewer successes in Group A (that the observer with a vested interest would be detrimental to the subjects' performance), so we will calculate the p-value as $P(X \leq 3)$.

(d) To determine the probability of 3 or fewer successes in Group A, add these four probabilities together. (This is the p-value.) How does this p-value compare to the approximate p-value from your simulation results from Investigation 1-9?

Recall that this probability is based on the assumption that randomization is the only factor at work; i.e., that the observer with the vested interest has no effect on subjects' performance in the video game.

(e) Discuss whether this probability suggests that the experimental results are very unlikely to have occurred by chance alone if there were no effect of the observer with the vested interest.

Discussion: You should have found the p-value of this test to be .0498, which should agree closely with your earlier simulation results.

$$\frac{C(11,3)C(13,9)}{C(24,12)} + \frac{C(11,2)C(13,10)}{C(24,12)} + \frac{C(11,1)C(13,11)}{C(24,12)} + \frac{C(11,0)C(13,12)}{C(24,12)} = .0498$$

This means that even if there were no genuine difference in the response variable between the groups, there's a little less than a 5% chance that randomization would produce a result favoring Group B subjects by as much as the experiment did. Since this probability is fairly small, we can conclude that the experiment provides moderately strong evidence in support of the researchers' conjecture (that the observer with the vested interest has a detrimental effect on subjects' performance).

This procedure is known as *Fisher's Exact Test*. It uses hypergeometric probabilities to determine the p-value for testing whether the difference in conditional proportions in a 2×2 table is statistically significant. The p-value is calculated as the probability of getting an experimental result as or more extreme, in the direction conjectured by the researchers, than the one actually obtained.

Calculation Details: You will obtain the same probabilities no matter which outcome you label as success and failure, no matter which explanatory variable group you focus on, and even regardless which variable you treat as the explanatory variable and the response, as long as you are consistent in your designations.

In the above table, we could have found the p-value as the probability of obtaining *8 or more successes* in Group B:

$P(Y{\geq}8)=$

Or the probability of 9 or more "non-threshold beaters" in Group A:

$P(Z{\geq}9)=$

Minitab Detour: Recall that you can use Minitab to perform these calculations:
Choose Calc > Probability Distributions > Hypergeometric to specify N, n, and the number of successes r (called X by Minitab).

Population size (N):	24
Successes in population (M):	11
Sample size (n):	12

You can tell Minitab to calculate the probability for an individual value, e.g., $P(X=3)$ using the input constant box:

⊙ Input constant:	3
Optional storage:	

You can also have Minitab calculate the *cumulative probability* for you, e.g., $P(X{\leq}3)$ to get the p-value direction. Select the cumulative probability button at the top of the window.

⊙ Cumulative probability

To see the entire probability distribution, enter the possible values of x into a column, say C1, and tell Minitab to store the results in an empty column.

⊙ Input column:	c1
Optional storage:	'prob'

To put many values into a column, you can type them in directly or you can use the "set" command, e.g.,

```
MTB> set c1      #This will place the values 0, 1,2, ..., 10, 11 into the column.
DATA> 0:11       # You can do this up to the number of successes in your collection
DATA> end        # to see the entire probability distribution.
```

Note, for calculating $P(Y \geq 8)$ as above, if you place 8 in the "input constant" box, Minitab will report $P(Y \leq 8)$. The cumulative probability in Minitab is always the probability for that value or less. We can obtain a "greater than probability" using the complement rule.

Probability Rule: The *complement rule* for a random variable says that $P(X \leq x) = 1 - P(X > x)$. In general, the *complement* of event A is the set of outcomes from the sample space that are not in the event of interest, often denoted by A^c. Since the probabilities of all possible outcomes must total one, $P(A^c) = 1 - P(A)$. Often times the complement rule comes in handy when the probability of the complement of the event is much easier to calculate.

So the complement of $P(Y \geq 8)$ is $P(Y < 8)$. With *discrete random variables*, it is important to note that $P(Y < 8)$ is *not* equal to $P(Y \leq 8)$. The events "$Y \leq y$" and "$Y \geq y$" are not complements since the outcome $Y = y$ is in both events. Since Y only takes on integer values, we must use the next lowest integer, $P(Y \leq 7)$. Thus to find $P(Y \geq 8)$ we will use $1 - P(Y \leq 7)$.

In general, for integer random variables, a very useful expression is $P(X \geq x) = 1 - P(X \leq x-1)$.

EFFECT OF SAMPLE SIZE

Now suppose that the psychology experiment had been conducted with twice as many subjects and that the results had turned out with identical proportions of success in both groups as in the actual study. In other words, suppose that 6 of 24 subjects in Group A had beaten the threshold, compared to 16 of 24 subjects in Group B.

(f) Create a two-way table to present these results.

(g) Verify that the proportions of success in the two groups are identical to your earlier analysis.

(h) How would a segmented bar graph for these data differ (if at all) from the one you created earlier? Explain.

(i) Before you conduct Fisher's exact test on these data, make a prediction as to whether the p-value computed here will be larger than before, smaller than before, or the same as before. Explain your reasoning.

(j) Now go ahead and conduct Fisher's exact test on these data. [Feel free to use Minitab.] Report the p-value, and show the details of your calculation. Also comment on whether the data provide strong evidence in support of the researchers' claim (that the observer in Group A causes subjects to perform worse on the task than the observer in Group B). Explain the reasoning process by which your conclusion follows from the p-value of the test.

(k) How does this p-value compare to the one from the original data? What does this say about the strength of evidence for the researchers' conjecture?

Discussion: You should have found that the p-value decreased from .0498 to .0042 when the sample size was doubled. The difference between 25% success in Group A and 67% success in Group B is more significant if there were 24 people in each group than if there only 12 people in each group. In other words, random assignment alone would produce such a difference between the groups more often with the smaller sample size than with the larger sample size. Taken to a further degree, if there were 100 people in each group, then it would be almost impossible for random assignment alone to produce a difference in success percentages as extreme as 25% vs. 67%. If the sample results had been found with that large sample size, then the test would produce an extremely small p-value, providing extremely strong evidence that the observer with a vested interest had a detrimental effect on subjects' performance.

Investigation 1-13: Minority Baseball Coaches

Recall from the Practice Problem at the end of Section 2 that of the 60 base coaches in Major League Baseball in 1999, 21 were minorities. Only 6 of those 21 minorities coached at third base, so the other 15 minority coaches worked at first base. (Therefore, 24 non-minorities coached at third base and 15 coached at first base.) Tolan argued that since third base is the more challenging position and typically leads to more managing responsibilities, this discrepancy constitutes evidence of discrimination against minority coaches.

(a) Use hypergeometric probabilities to perform Fisher's exact test, assessing the degree to which these data suggest that the difference in proportions coaching at third base is statistically significant. Report the p-value of the test, along with a description of how it is calculated. Write a paragraph or two reporting on your findings and explaining how your conclusions follow from your analysis.

(b) Is this a controlled experiment or an observational study? Explain your answer. What is the implication of this study design on whether you can draw a cause-and-effect conclusion from this study? Explain.

Discussion: Of the 21 minority coaches, only 28.6% (6/21) were at third base, compared to 61.5% (24/39) of the non-minority coaches. Let X represent the number of minority coaches at third base. If we were to randomly assign the 30 third base coaching jobs among the 21 minorities and 39 non-minorities, the probability of getting only 6 or fewer minorities assigned to third base is:

$$P(X \leq 6) = \frac{\binom{21}{6}\binom{39}{24}}{\binom{60}{30}} + \frac{\binom{21}{5}\binom{39}{25}}{\binom{60}{30}} + \ldots + \frac{\binom{21}{0}\binom{39}{30}}{\binom{60}{30}} = .015$$

This is a small p-value, indicating that such an extreme result would happen in only about 1.5% of all random assignments in the long run. Thus, the difference in the response between the two explanatory variable groups is *statistically significant*, and we have fairly strong evidence that if there was no difference between the 2 groups that random chance alone would not have produced such a large discrepancy in third base assignments between the minority coaches and the non-minorities.

But, it is important to note that this is an observational study and not a controlled experiment, because the researcher did not impose the treatment (being a minority or not) on the coaches. Thus, we can not draw a cause-and-effect conclusion connecting the coaches' ethnicity to their coaching assignments. For example, perhaps seniority and experience are confounding variables (since most older coaches are not minority coaches).

In fact, some would question whether it's even valid to conduct a test and produce a p-value in this situation. For the psychology experiment, we were on firm footing by conducting this test, because we were taking into account the random assignment of subjects to groups that the researchers actually used in the experiment. But with this baseball coaching study, no randomization was used, only observations were made. All statisticians would agree with the correctness of the p-value calculation and with the caution that we not draw a cause-and-effect conclusion, but some would argue that without randomization in the actual study, it is inappropriate to perform this test and calculation. Our view is that as long as you are clear about what you are calculating (.015 is the probability of getting so few minorities assigned to third base *if randomization had been* the method of allocation), and as long as you do not draw a causal conclusion from an observational study, the test does shed some light into how "significant" that difference between 28.6% (6/21) and 61.5% (24/39) really is.

Practice:

1-22) Relieving Back Pain
Recall the study of the drug botulinum from Practice 1-19, in which researchers examined whether the drug botulinum toxin A is helpful for reducing pain among patients who suffer from chronic lower back pain. Fifteen subjects were randomly assigned to use the drug and the other 16 subjects received saline. Nine of the 15 subjects receiving the drug experienced a substantial reduction in pain, compared to 2 of the 16 subjects in the saline group.
(a) Calculate the p-value from Fisher's exact test for this study. [*Hint*: Note that the researchers conjectured that the botulinum group would experience a *higher* proportion of pain reduction, so you want to find the probability of getting 9 or more "successes" in that group if in fact there was no difference between the treatments and so the 11 who experienced pain reduction would have done so regardless of which treatment they received.]
(b) Write a paragraph describing the conclusion that you would draw from the study, based on results of your analysis, to the researchers. In addition to stating your conclusion, explain the reasoning process that leads to this conclusion (What does this p-value tell you?).
(c) Does the design of this study allow you to draw a cause-and-effect conclusion regarding botulinum and pain reduction? Explain.

1-23) Gender Discrimination
In a now famous study of gender discrimination, Rosen and Jerdee (1974) report on a study of 48 male bank supervisors (attending a management institute at the University of North Carolina) who were each sent the same personnel file and asked to judge whether the person should be promoted to a branch manager position. The files were identical except that half of them were identified as the file of a female and the other half indicated that the file was that of a male. The researchers suspected that a higher proportion of "males" would be recommended for promotion. Of the 24 "male" files, 21 were recommended for promotion. Of the 24 "female" files, 14 were recommended for promotion.
(a) Is this an experiment or an observational study? Explain.
(b) In this study, the subjects believed that they were all participating in an identical exercise dealing with personnel problems in the banking industry. Explain why it is important for the subjects to be "blind" to the fact that there were other versions and that the researchers were focusing on the gender of the applicant.
(c) Calculate the p-value from Fisher's exact test for this study.
(d) Write a paragraph describing the results of your analysis to the researchers. In addition to stating your conclusion, explain the reasoning process that leads to this conclusion (What does this p-value tell you?).
(e) Does the design of this study allow you to draw a cause-and-effect conclusion regarding gender and likelihood of promotion? Justify your answer.

1-24) Comparing p-values
In the Minority Baseball Coaches study, the p-value was .015, and in the Friendly Observer's study, the p-value was .0498.
(a) Which study provides stronger evidence that the difference between the groups was larger than we would expect by chance? Explain.
(b) Which study provides stronger evidence that the difference in the response was caused by the explanatory variable? Explain.

CHAPTER SUMMARY

This chapter has explored the analysis of two-way tables and the types of conclusions that can be drawn from different studies. The main distinction to keep in mind is the difference between *observational studies* and *experiments*. In experiments, such as the Friendly Observers study and the tripping experiment, the researchers impose a treatment on the subjects in two or more groups. In observational studies, such as case-control studies, cohort studies, and cross-sectional designs, the researchers do not impose a treatment (as in the smoking studies and the Minority Coaches study). With observational studies, there is always the possibility of *confounding* variables which prevents us from being able to draw cause-and-effect conclusions, no matter how large the difference in proportions or how large the *odds ratio* is. This was reinforced by the discussion of *Simpson's paradox* which showed how a third variable can invalidate, or even reverse, the direction of the association between two variables. On the other hand, by randomly assigning the subjects to the treatment groups in an experiment, we control for such confounding variables by creating groups that should be similar prior to the imposition of the treatment.

A second huge advantage of randomly assigning the subjects to treatment groups is that we can determine the likelihood of different response results assuming there is actually no difference between the treatment groups. This helps us judge when an observed difference is large enough to convince us that it did not happen by chance alone and instead provides evidence of a *statistically significant* treatment effect. We first used simulation and then a probability distribution (*hypergeometric*) to assess how often different results happened just by chance. This *p-value* (the *probability* of observing an outcome at least as extreme as that obtained in the study in the direction conjectured by the researches) indicates how strong the evidence is against the assumption that there is no treatment effect. We saw that these p-values are affected by sample size and that we need to exert some caution in interpreting them. Our conclusions should always consider the study design that produced the data. Even a p-value close to zero does not allow us to draw a cause-and-effect conclusion with an observational study. It is also important to consider other design issues: Was the design retrospective so that there could be errors arising from people's memory and perceptions? Were the subjects blind to the treatment being imposed so that did not affect their reactions or the researchers' measurements of their responses?

The studies in this chapter all involved a categorical response variable and a categorical explanatory variable. In this setting, data are often organized in a two-way table. In analyzing two-way tables, we recommend using *segmented bar graphs* to visually compare the outcomes of the response variable for two or more groups and *difference in proportions, relative risk,* and *odds ratios* to assess the strength of the association numerically. It is often more useful to report the relative risk and/or the odds ratio, especially when the baseline proportion/risk of "success" is small. Moreover, the difference in proportions and relative risk are not reasonable summaries to use with case-control studies in which the researcher has determined the distribution of the response variable by the design of the study.

In the next chapter we will follow the same progression of describing appropriate numerical and graphical summaries and then determining if there is a statistically significant difference between groups, but for *quantitative* response variables. It will be important to keep in mind that we always need to consider how the data were collected in order to determine the scope of conclusions we are able to draw. You will also continue to practice the first step: identifying the variables and observational units clearly.

TECHNOLOGY SUMMARY

- In this chapter you learned how to use Excel to create segmented bar graphs and to calculate combinations, as well as for more basic calculations such as relative risks and odds ratios.
- You also learned several useful things about Minitab along the way in this chapter:
    - o Simulation of random assignment of the explanatory variable and then *unstacking* the observational units into these randomly assigned explanatory variable groups.
    - o Using a macro to repeat the same set of Minitab commands many times. Remember to store the results from each repetition using a *counter* (e.g., k1) and to initialize that counter to 1 before you start the macro.
    - o How to create an *indicator variable* of 0's and 1's to correspond to certain events or relationships occurring.
    - o Creating a list of sequential values using the *set* command.
    - o Performing algebraic manipulation using the *let* command.
    - o Calculating *hypergeometric probabilities* using the probability and the *cumulative probability* buttons. Also creating and plotting an entire *probability distribution* and determining the *expected value* using this distribution.
- You also used some java applets to explore the important concepts of randomization and statistical significance. These applets helped you visualize the long-run patterns that we expect to see by conducting many more repetitions than we would want to carry out by hand. You may want to continue to use the version of the Friendly Observers applet that allowed you to change the table entries to carry out simulations of Fisher's Exact Test though now you also have the hypergeometric probability distribution to calculate the p-values exactly.

CHAPTER 2: COMPARISONS WITH QUANTITATIVE VARIABLES

This chapter parallels the first one in many ways. We will continue to consider studies where the goal is to compare a response variable between two (or more) groups. The difference here is that these studies will involve a *quantitative* response variable rather than a *categorical* one. The methods that we employ to analyze these data will therefore be different, but you will find that the basic concepts and principles that you learned in chapter 1 still apply. These include the principle of starting with numerical and graphical summaries to explore the data, the concept of statistical significance in determining whether the difference in the distribution of the response variable between the two groups is larger than we would reasonably expect from randomization alone, and the importance of considering how the data were collected in determining the scope of conclusions one can draw from the study.

SECTION 2-1: SUMMARIZING QUANTITATIVE DATA

In this section, we will begin our exploration of numerical and graphical summaries when the response variable is quantitative and we want to compare two or more groups. You will learn a new set of techniques, but keep in mind that the overall goal is to summarize the data informatively and provide effective comparisons of the distributions. First you will practice developing some intuition for how you might expect some categorical and quantitative variables to behave. Then you will focus on more descriptive (graphical and numerical) tools for comparing the distributions of a quantitative variable between groups. Finally, just as in chapter 1, you will simulate randomization distributions to judge when the observed difference between two groups is large enough to be considered statistically significant.

Investigation 2-1: Anticipating Variable Behavior

One skill that you will develop as you look at more and more graphs is being able to anticipate variable behavior. You should have intuition for how you expect a variable to behave and what properties you expect the distribution of its values to have. This will also help you know when to be surprised that a variable is not behaving as expected or when a distribution looks unusual and should be investigated further.

The following activity will help you develop this intuition by asking you to identify which variables most plausibly lead to which graphs. An introductory statistics class of 35 students ("Probability and Statistics for Engineers and Scientists") was given a survey on the first day of class. The survey questions were:
     a) To the nearest inch, how tall are you (in inches)?
     b) Which do you prefer, Coke or Pepsi?
     c) How many siblings (brothers and sisters) do you have?
     d) How much did you pay (including tip) for your last haircut?
     e) Are you male or female?
     f) Guess your professor's age in years
Their responses are displayed in the graphs below. We have purposely deleted the axis and scale labels for both axes, but we can tell you that the variable is displayed along the horizontal axis and the height of each bar represents the number of observations at that outcome or in that range of values. Your task is to match each graph to one of the above variables. You will be asked to provide a coherent argument for why the graph selected makes sense for the variable. Keep in mind, there is not necessarily only one believable assignment. What is most important is being able to justify your choices.

Write a paragraph explaining how you matched graphs. For example, what features helped you decide and how did you distinguish between the graphs that were similar?

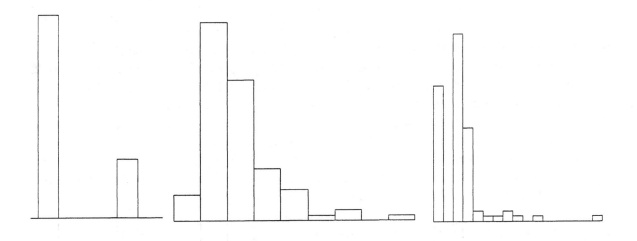

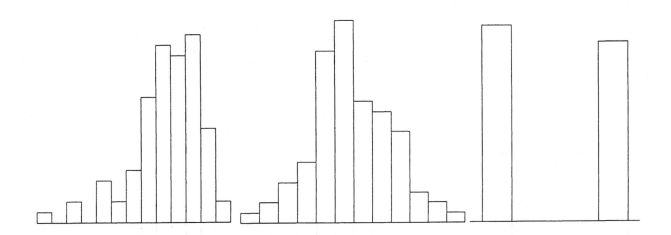

Discussion:

One consideration in thinking about how variables will behave is to consider whether the variable is quantitative or categorical. In this activity, "favorite soda" and "gender" are the only categorical variables and therefore correspond to the graphs ("bar graphs") with only two possible outcomes. However, it is not so easy to know which is which. Was this a "Pepsi campus"? If it were, would that make people more or less likely to pick one soda as their favorite? What is the gender breakdown at this university? In this class? We do have some clues in that this was a class for scientists and engineers—how might that impact the gender breakdown? We are less concerned that you know the answers to all these questions, but that you have considered them and have justified your choice. Similarly, graphs 2 and 3 are similar in that they both have many observations with small values and not as many observations with larger values. Which variables did you match those up with? How did you decide between them? Based on what you know about "biological measurements," how do you expect the height variable to behave? Knowing that the students were pretty much guessing at random of the instructor's age (though having seen the instructor on the first day), how do you think those guesses will behave?

We are resisting telling you "the answers" here since in much of statistical analyses, we never know "for sure." But we have evidence and we can support our arguments based on that evidence. Then it is our responsibility to try to convince the reader that we have considered the strength of the evidence and drawn a reasonable conclusion. Often, different people will come to different reasonable conclusions. It is up to you to look through the fancy graphs and prose in order to see what the data tell you!

As you learn techniques for summarizing the behavior of these variables, keep in mind that your "job" is to convey the overall behavior of the variable to the reader. In particular, remember to always relate your comments to the context. Never again in this class should you look at graphs without knowing which variables are being represented.

Investigation 2-2: Cloud Seeding

Our lives depend on rainfall. Consequently, scientists have long investigated whether humans can intervene and, as needed, help nature produce more of rainfall. In one study researchers in southern Florida explored whether injecting silver iodide into cumulus clouds would lead to increased rainfall. On each of 52 days that were judged to be suitable for cloud seeding, a target cloud was identified and a plane flew through the target cloud in order to seed it. Randomization was used to determine whether or not to load a seeding mechanism and seed the target cloud with silver iodide on that day. Radar was used to measure the volume of rainfall from the selected cloud during the next 24 hours. The results (measured in volume units of acre-feet, "height" of rain across one acre) are presented below (from Simpson, Alsen, and Eden, 1975):

Unseeded:

1.0	4.9	4.9	11.5	17.3	21.7	24.4	26.1	26.3	28.6	29.0	36.6	41.1
47.3	68.5	81.2	87.0	95.0	147.8	163.0	244.3	321.2	345.5	372.4	830.1	1202.6

Seeded:

4.1	7.7	17.5	31.4	32.7	40.6	92.4	115.3	118.3	119.0	129.6	198.6	200.7
242.5	255.0	274.7	274.7	302.8	334.1	430.0	489.1	703.4	978.0	1656.0	1697.8	2745.6

(a) Is this an experiment or an observational study? Explain. Also identify the observational/experimental units in this study.

experimental study

(b) Identify the explanatory variable (EV) and the response variable (RV). Also indicate what type of variable (categorical or quantitative) each is.

EV: *unseeded or not*   type: *binary categorical*

RV: *volume of rainfall*   type: *quantitative*

Include a schematic of the experimental design:

Clouds → seeded → rainfall
↘ unseeded
randomization

(c) Explain why randomization was used in this study.

(d) What do you think is the benefit of not telling the pilot and the experimenters whether the plane's seeding mechanism was loaded?

Because the response variable in this study (rainfall amount) is quantitative, we cannot use the techniques presented in chapter 1 to analyze and draw conclusions from these data. We will be able to stick to the basic principles that you learned earlier, though. For example, we start to analyze the data by constructing and interpreting graphical displays.

The simplest graphical display for comparing the distributions of a quantitative response variable between groups is the dotplot, as you encountered in chapter 1. Recall that a variable's distribution refers to its pattern of variation. Especially important when using dotplots to compare groups is that *parallel dotplots* be drawn on the same scale. Dotplots of these rainfall amounts are presented below:

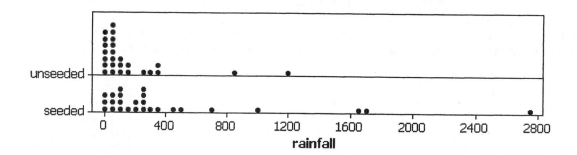

(e) Write a few sentences to compare and contrast the distributions of rainfall amounts between these two groups of seeded and unseeded clouds. Based on these observations, would you say there was evidence that the clouds that were seeded had substantially more rainfall?

rainfall heppens @ the low level.
The Centr. the Seeded has more then unseeeled.

Discussion: An important idea when comparing distributions between two or more groups is that of a *statistical tendency*. In this study rainfall amounts *tended* to be higher for seeded clouds than for unseeded clouds. The rainfall amounts from the seeded clouds appear to be centered around 200-400 acre-feet, whereas the rainfall amounts from unseeded clouds appear to be centered around 50-150 acre-feet. A tendency is not a hard-and-fast rule: There is a lot of overlap between the two distributions, and there are many "unseeded" measurements that are larger than many "seeded" measurements. But *in general*, or *on average*, the seeded clouds *tended* to produce higher rainfall amounts than the unseeded clouds. Whether this difference is large enough to be statistically significant (i.e., unlikely to have happened by chance) is a different issue that we will return to later in this chapter.

The center of the distribution is the most important feature to compare in this case, because the goal of the study was to see if cloud seeding tended to increase rainfall amounts. We notice some other features in the dotplots, as well. The rainfall amounts of the seeded clouds are more spread out than those for the unseeded clouds. Both distributions are *skewed to the right*, which means that there are many values at the low end of the scale and fewer values tailing off at the high end. [See sketches below.] Both distributions have some unusually large values: 830.1 and 1202.6 acre-feet for the unseeded group, 1656.0, 1697.8, and 2745.6 acre-feet for the seeded group.

In general, some features to look for when commenting on the distribution of a quantitative variable include:

- *Center*: Around what value is the distribution centered? Approximately where is the middle value taken on by the variable?
- *Spread*: How much spread or variability does the distribution have? Are most values close to the center or are most of the values far from the center?
- *Shape*: Is the distribution symmetric or skewed? Does it have one peak or several? Are there gaps in the distribution? See four common shapes in the following dotplots.
- *Unusual observations*: Are there any observations that do not follow the same general pattern as the rest of the data? In particular, are there extreme values that are located far above or below the majority of the data (*outliers*)?

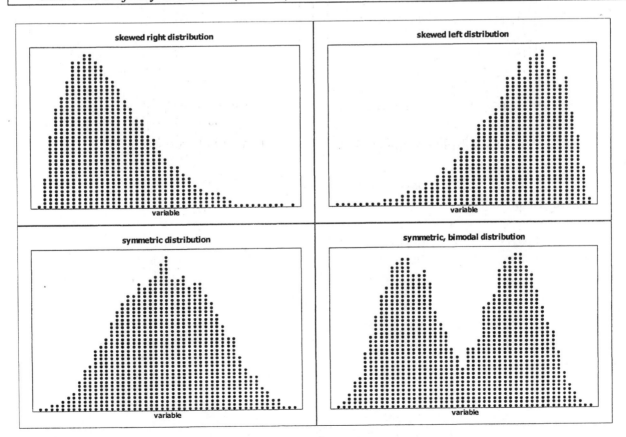

Dotplots can be very informative because they contain all of the original data, but providing a numerical *summary* of the data can also be very helpful, especially when making comparisons between groups and examining larger data sets.

Def: The *five-number summary* "reduces" the distribution of data to these values:

- Minimum
- First (or lower) quartile, also known as the 25th percentile, a value that 25% of the data falls below and 75% above (after the data have been arranged in order), denoted by Q_1.
- Median, or 50th percentile, the middle value when the data have been arranged in order so that that half of the data falls below this value and half above.
- Third (or upper) quartile, or 75th percentile, a value that 75% of the data falls below and 25% above (after the data have been arranged in order), denoted by Q_3.
- Maximum

Calculating these summaries can get a bit tricky, because it often happens that no value has exactly 25% below and 75% above it. When we calculate these summaries by hand, we will use the following algorithms:

- When the number of observations, *n*, is odd, the median is the unique value in the middle. It can be found in position $(n+1)/2$ once the data are arranged in order.
- When there is an *even* number of observations, we will take the median to be the average of the middle two, which can be found in positions $n/2$ and $(n/2)+1$.
- We will specify the *first* quartile Q_1 as the median of the values *below* the *location* of the (overall) median.
- We will specify the *third* quartile Q_3 as the median of the values *above* the location of the (overall) median.
    - When there are an odd number of observations and the location of the median coincides with a specific value in the data set, we will not include the overall median in either half of the data for the purpose of finding quartiles. That is, we find the quartiles by looking at the values strictly above and strictly below the location of median. With an even number of observations, the location of the median falls between two values and we look at the observations above and below that location.

Be aware that different calculators and software packages calculate quartiles using slightly different algorithms.

(f) Calculate the five-number summary of rainfall amounts for both groups. Record your results below. The data are reproduced here for your convenience; note that the sample size is 26 in each group and that the values are already in order:

Unseeded:

1.0	4.9	4.9	11.5	17.3	21.7	24.4	26.1	26.3	28.6	29.0	36.6	41.1
47.3	68.5	81.2	87.0	95.0	147.8	163.0	244.3	321.2	345.5	372.4	830.1	1202.6

Seeded:

4.1	7.7	17.5	31.4	32.7	40.6	92.4	115.3	118.3	119.0	129.6	198.6	200.7
242.5	255.0	274.7	274.7	302.8	334.1	430.0	489.1	703.4	978.0	1656.0	1697.8	2745.6

	Minimum	Lower quartile	Median	Upper quartile	Maximum
Unseeded					
Seeded					

(g) Does comparing the five-number summaries reinforce the earlier conclusion that seeded clouds tend to have larger rainfall amounts than unseeded clouds? Explain.

fan Yiyi-y

The five-number summary can also be displayed graphically with a *boxplot*, also known as a *box-and-whiskers plot*. Below is the boxplot of rainfall amounts for the unseeded clouds:

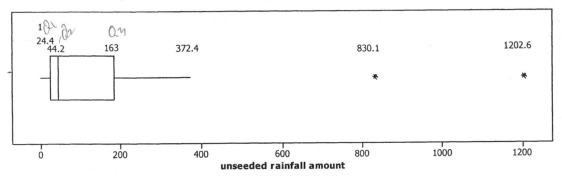

To construct a boxplot:
- Start with an axis scaled to include all of the values taken by the variable.
- Draw vertical lines at the quartiles, and connect those lines to form a box around the middle 50% of the data. Also draw a vertical line at the median.
- Mark outliers with an asterisk. Our primary rule for identifying outliers will be to find any observations that fall more than 1.5 × (the length of the box) away from the box. (The difference between the quartiles is called the *inter-quartile range*, or IQR. So the outlier boundaries are found by calculating $Q_1 - 1.5IQR$ and $Q_3 + 1.5IQR$. Any observations outside these boundaries are classified as outliers according to the *1.5IQR criterion*.)
- Extend horizontal lines (whiskers) from the boxes out to the most extreme non-outlier.

(h) Use the outlier rule described above to determine if there are such outliers in the "seeded" group. List the values for any outliers that you find. [*Hint:* Even though the dotplot suggests that there are probably no low outliers, be sure to check on the low end as well as the high end anyway.]

Comparative boxplots drawn on the same scale can be especially useful for comparing distributions of a quantitative variable between two groups.

(i) Construct a boxplot of rainfall amounts for seeded clouds on the same scale below:

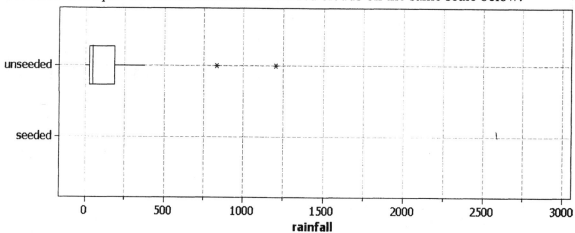

(j) Comment on what these boxplots reveal about the distributions of rainfall between the two seeding conditions.

Def: Another numerical measure of the center of a distribution is the *average* or *mean* of the distribution. Often denoted by $\bar{x}$, the mean is the arithmetic average, calculated by summing the data values $(x_1,\ldots, x_n)$ and dividing by the sample size, n: $\bar{x} = \dfrac{\sum\limits_{i=1}^{n} x_i}{n}$

(k) Which do you think will be larger: the mean rainfall for the seeded clouds or the mean rainfall for the unseeded clouds? Why?

(l) The mean rainfall turns out to be 164.6 acre-feet for the unseeded clouds and 442.0 acre-feet for the seeded clouds. Mark these values with an 'x' on the boxplots in (i). How do these mean values compare to the median values that you calculated above?

(m) How many and what proportion of the unseeded clouds have rainfall amounts that exceed the mean (164.6)? Answer the same question for the seeded clouds also. [*Hint*: Look back at the original data and/or at the dotplot.]

(n) Based on your answer to (m), would you say that the mean values effectively summarize a "typical" rainfall amount for each group? Explain.

With a skewed distribution, the mean is typically pulled in the direction of the longer tail. Thus, with a skewed right distribution, the mean will tend to be larger than the median. In this situation (skewed data), the median does a better job of describing a "typical" value of the distribution since it has an equal number of observations on each side. On the other hand, because the mean considers the *magnitudes* of all values in the data set, it can be much larger (or smaller) than most of the data.

Study Conclusions: These graphs and summary statistics suggest that the seeded clouds did in fact tend to produce more rainfall than the unseeded clouds. All of the values in the five-number summary are higher for the seeded clouds, with the median higher by 222-44.2=177.8 acre-feet of rain. Similarly, the mean rainfall for the seeded clouds is larger by 277.4 acre-feet. We would need to do a more formal analysis to determine if this difference is unlikely to have occurred just through randomization. If it is, then we would be justified in attributing the increase in rainfall to the cloud seeding since this study was a controlled, randomized experiment. The individuals doing the cloud seeding and measuring the amount of rainfall were also blind to whether or not the silver iodide had been sprayed into each cloud.

Minitab Detour:

Minitab can perform these calculations and graphs for us as well, though you will see that Minitab calculates quartiles with a slightly different algorithm. Excel can also perform these calculations and graphs but many statisticians do not find Excel's statistical capabilities as effective as a standard statistical package such as Minitab.

In Minitab, select File > Open Worksheet and open the CloudSeeding.mtw worksheet. Typically each row in the Minitab worksheet represents an observational unit, and each column a variable. In this worksheet the rainfall variable is stored in Column 1 and the explanatory variable is stored in Column 2. You will see that Column 2 is denoted as a *text column* by the T next to the C2 at the top of the column.

- To create boxplots on the same scale as you did in the above investigation, choose Graph > Boxplot and choose the "With Groups" option. Click OK. Double click on C1 in the left window and it will appear in the Graph variables box (or just type C1or rainfall in the Graph variables box). Then click in the Categorical variables box and type C2. Click the Scale button and check the "Transpose value and category scales" box (this will give the boxplots the same orientation as in the above investigation). Click OK twice.

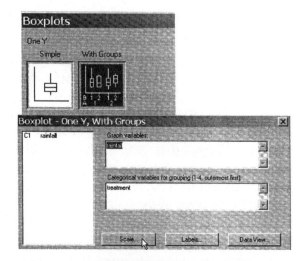

- To create parallel dotplots, choose Graph > Dotplot and select the "With Groups" option. Enter the rainfall variable in the Graph variables box and the treatment variable in the Categorical variables box. Click OK.

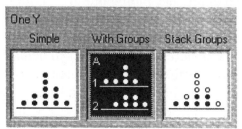

- To obtain descriptive statistics for each group, choose Stat > Basic Statistics > Display Descriptive Statistics. Again, enter the rainfall variable in the Variables box and then click in the By variable box and enter the treatment variable. Click the Statistics button and check the Mean, First quartile, Median, Third quartile, Interquartile range, Minimum, and Maximum boxes. (Unselect the other boxes.) Click OK twice. You will notice that Minitab's quartiles are slightly different than those we calculate by hand (which can affect the 1.5IQR test for outliers).

Variable	treatment	Mean	Minimum	Q1	Median	Q3	Maximum	IQR
rainfall	seeded	442	4.10	79.5	222	445	2746	365
	unseeded	164.6	1.00	23.7	44.2	183.3	1202.6	159.6

To copy and paste this output, you might want to try Courier 10pt font.

- Recall that you can *unstack* the columns so that each group's data values are in their own column. The easiest way to do this is to type at the MTB prompt (make sure you have enabled commands).

```
MTB> unstack c1 c3 c4;
SUBC> subs c2.
```

You can also stack the columns if they came to you in the unstacked format:

```
MTB> stack c3 c4 c5;          #stacks c3 above c4, storing the
SUBS> subs c6.                #results in C5 and C6.
```

After running this command, C6 will be an *indicator variable*, using a 1 if the data are from the first column (C3 here) and a 2 if the data came from the second column (C4 here). You can also use the usenames subcommand to use the category names in C6. The corresponding menu commands for these procedures can be found at Data > Unstack and Data > Stack.

TRANSFORMATIONS

The researchers' goal in the cloud seeding study was to compare the average or typical amount of rainfall under the two treatment conditions. This comparison can be difficult if the shape and spread of the distributions are quite different. Often, such comparisons are easier to make after the data have been *transformed*. Transformations to make the distribution more symmetric will also be useful later in the course for applying certain inference procedures, and transformations are also helpful for fitting certain statistical models to data.

(o) In Minitab, take the logarithm of the rainfall data:
```
MTB> let c3=log(c1)  #Minitab assumes natural log
```
[*Habit*: Make sure you name new columns after you create them!]

Now create parallel dotplots and boxplots, and calculate the same descriptive statistics for this transformed data.

(p) Are the spread and shape of the distributions more similar now? Explain.

(q) Does one group display a tendency for higher values? Explain.

Discussion: The log transformation is often helpful for "bringing in the tail" of a distribution that is skewed to the right. By having more symmetric distributions, with similar spreads, you can often make a more direct comparison of the centers. One advantage to using the log function is that it is *monotonic* so it preserves the ordering of the data values. Thus, noting that log(rainfall) tends to be higher in the seeded clouds than in the unseeded clouds leads us to infer that the rainfall amounts themselves also have a tendency to be larger in the seeded clouds

In addition to the log transformation, other common choices are power transformations that simply raise the data values to a power. Powers less than one, such as square root and reciprocal, are effective with distributions that are skewed to the right, while powers greater than one, such as square and cube, are helpful with distributions that are skewed to the left.

Practice:

2-1) Roller Coaster Speeds
The Roller Coaster Database maintains a
web site (www.rcdb.com) with data on roller
coasters around the world. Some of the data
recorded include whether the coaster is
made of wood or steel and the maximum
speed achieved by the coaster, in miles per
hour. The boxplots display the distributions
of speed by type of coaster for 145 coasters
in the United States, as downloaded from the
site in November of 2003.

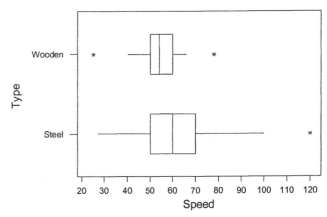

(a) Identify the observational units in this study. Then identify the explanatory and the response
variable here. Also indicate for each whether it is quantitative or categorical.
(b) Summarize what these boxplots reveal about the differences between the speeds of wooden
and steel roller coasters. In particular, is there a tendency for one type of coaster to be faster?
Explain.
(c) Do these boxplots allow you to determine whether there are more wooden or steel roller
coasters?
(d) Do these boxplots allow you to say which type has a higher percentage of coasters that go
faster than 60mph? Explain and, if so, answer the question.
(e) Do these boxplots allow you to say which type has a higher percentage of coasters that go
faster than 50mph? Explain and, if so, answer the question.
(f) Do these boxplots allow you to say which type has a higher percentage of coasters that go
faster than 48mph? Explain and, if so, answer the question.
(g) Write a paragraph comparing and contrasting these distributions. Describe the shape, center,
and spread (as best you can) for each distribution, and then also comment on the issue of whether
one type of coaster tends to have higher speeds than the other. Remember to state your
description in the context of the study.

2-2) More on Roller Coasters
(a) Open the Minitab worksheet `coasters.mtw`, which contains data on 145 coasters in the
United States, as downloaded from the www.rcdb.com site in November of 2003. Use Minitab
to produce boxplots of height (in feet) by type, length (in feet) by type, and drop (in feet) by
type. Write a paragraph summarizing differences between wooden and steel coasters with regard
to these variables.
(b) Another variable in the file is "age group" (C13), which is coded as "1:older" for coasters
opened in 1990 or earlier, coded as "2:middle" for coasters opened between 1991 and 1998
inclusive, and coded as "3:newer" for coasters opened in 1999 or later. Produce boxplots of
height, length, drop, and speed by this age group variable. Write a paragraph summarizing how
roller coasters appear to have changed over time with respect to these variables.

Investigation 2-3: Geyser Eruptions

Millions of people from around the world flock to Yellowstone Park in order to watch eruptions of Old Faithful geyser. How long does a person usually have to wait between eruptions, and has the timing changed over the years? In particular, scientists have investigated whether a 1998 earthquake lengthened the time between eruptions at Old Faithful. The data in oldfaithful.mtw are the intereruption times (in minutes) for all eruptions occurring between 6am and midnight on August 1-8 in 1978 (from Weisberg (1985) and http://www.geyserstudy.org/OldFaithful_data.htm) and also in 2003.

(a) Is this an experiment or an observational study? Explain.

(b) Open the oldfaithful.mtw worksheet. Since each variable is in a separate column, choose Graph > Boxplot and choose the "Multiple Y's, Simple" option. Enter the 1978 data (C1) and a comparable set of data from 2003 (C3) into the "Graph variables" box. Click the Data View button and check the "Mean symbol" box. Click OK twice. Write a paragraph comparing the two distributions. In particular, is there evidence of a tendency for longer intereruption times in one year? Justify your conclusion.

(c) Park rangers and visitors would also appreciate if Old Faithful was as reliable as its name implies. Is there evidence of a tendency for the intereruption times to be more *consistent* in one of these two years? Justify your conclusion.

In some cases comparing the *variability* of the group distributions is as interesting or even more interesting than comparing the centers. One simple way to measure the variability (spread) of a distribution numerically is to calculate the *range*, which is the difference between the maximum and minimum values in the data set.

(d) Calculate the range in the intereruption times for each year. [You should be reporting a single number for each year.]

    1978:   2003:

(e) Suppose the two smallest intereruption times in 2003 were removed from the data set and the range was recalculated. Would this new value be appreciably different from the previous value?

(f) Do you think the range is a particularly useful measure of spread? Explain.

Discussion: The range is a widely reported measure due to its simplicity. Keep in mind when we refer to "range" we are meaning for you to calculate this individual number (as opposed to saying 56-110 minutes). A severe limitation is that the range really just tells us about two values in the distribution. Other measures of spread do a much better job of summarizing the spread throughout the distribution.

Def: The *inter-quartile range* (IQR) is the difference between the third quartile and the first quartile. The IQR is one number measuring the width of the middle 50% of the data values and corresponds to the width of the box segment of the boxplot.

(g) Use the boxplots to estimate as well as possible the inter-quartile range of Old Faithful's intereruption times in each year. [Holding the mouse over the box should reveal the quartile values and IQR.] In which year was the IQR smaller?

    1978:   2003:   smaller:

(h) Does the year with the smaller IQR agree with your answer to (c) about which year has more "consistent" intereruption times? In other words, are distributions with smaller "spread" characterized by larger values of the IQR or smaller values of the IQR? Explain.

Another measure of spread examines the deviations of the data values from the mean. Summing the deviations from the mean will not be informative since that sum is always zero! Summing the absolute deviations from the mean provides a positive sum. We can also sum the squared deviations from the mean. To describe a "typical" squared deviation we will average these values by dividing by n-1.

$$\frac{\sum_{i=1}^{n}(x_i - \bar{x})^2}{n-1}$$

(i) What are the units of this quantity, often called the *variance* (e.g., if x is measured in minutes as with the intereruption times)?

Def: The *standard deviation* is the square root of the variance, often denoted by s:

$$s = \sqrt{\frac{\sum_{i=1}^{n}(x_i - \bar{x})^2}{n-1}}$$. It can loosely be interpreted as the size of a typical deviation from the mean of the data set. The square root gives us a value that has the same units as the variable. We will use technology to calculate this value.

(j) In Minitab, select Stat > Basic Statistics > Display Descriptive Statistics and enter C1 and C3 into the Variables box and click OK. Minitab will include the standard deviation (StDev) among the numerical summaries. Report the standard deviation, with appropriate units, for each year. In which year was the standard deviation smaller?

    1978:   2003:   smaller?

(k) Does the year with the smaller standard deviation agree with your answer to (c) about which year has more "consistent" intereruption times? In other words, are distributions with smaller "spread" characterized by larger values of the standard deviation or smaller values of the standard deviation? Explain.

(l) Delete the two low intereruption times from the 2003 data (56 and 57 minutes). [*Hint*: Use the backspace key to delete these individual values from C3 in the Data window.] Use Minitab to recalculate the standard deviation and IQR. Compare these values to those determined in (g) and (j). Which value has changed more?

Discussion: The standard deviation is 8.46 minutes for the 2003 data, 12.97 minutes for the 1978 data. Note that the standard deviation assumes that the typical distance is the same above the mean as below the mean. This is less useful for skewed distributions. In particular, for skewed data or if outliers are present, the IQR is a better measure of spread to report than the standard deviation. Furthermore, if the two low intereruption times were deleted from the 2003 data the standard deviation changes to 6.87 minutes (the IQR does not change). The standard deviation will be affected by outliers (is not *resistant*) and so is not the best choice when there are extreme outliers in the data set. However, when the data are symmetric, the mean and standard deviation are the preferred numerical summaries to report. With skewed data, the median and IQR are preferred.

(m) Suppose that you arrive just as one eruption is ending, and you want to know how likely you are to have to wait less than an hour for the next eruption. Approximately how often did that happen in each year? What about for a wait of less than 75 minutes?

By their nature, boxplots reduce the data set to just 5 numbers (and may display any outliers if they are present). They quickly reveal where the boundaries of where the bottom 25%, next 25%, next 25%, and top 25% of the data fall, but they do not provide as much information about other aspects of the distribution. To recover other information about the distributions, particularly their shapes, other graphical summaries are more appropriate.

> **Def:** Another common and helpful visual display for quantitative variables is a *histogram*, which splits the range of values taken by the variable into intervals, counts how many of the observational units fall into each interval, and displays a rectangle for each interval whose height is proportional to that count. When comparing distributions, the histograms should be drawn on the same scale.

(n) In Minitab, choose Graph > Histogram and select the "Simple" option. Enter the 1978 (C1) and 2003 (C3) data into the variables box. Click the Labels button, then select the Data Labels tab and click the "Use y-value labels" button. Click OK. Click the Multiple Graphs button and check the "Same X" box. Click OK twice.

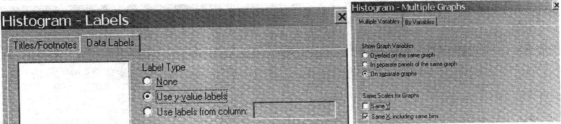

What percentage of intereruption times fall below 60 minutes for each year? What percentage fall below 75 minutes in each year?

(o) What else do these histograms reveal about the distributions of intereruption times that the boxplots did not reveal?

Histograms (and dotplots) are more useful than boxplots for examining the *shape* of a distribution. In particular, there may be clusters or gaps in the distribution that are not revealed by the boxplots.

Often we can get a better idea of the long-term pattern of a distribution by looking at a larger number of observations.

(p) The data in C4 contains the 2003 data on intereruption times through August 27, 2003, for both daylight and nighttime (current technology allows for measurements to be mechanically recorded). Produce a histogram of these data and describe the distribution. [As always, remember to make your statements in context.]

Study Conclusions: These data reveal that the distribution of intereruption times does seem to have changed substantially between 1978 and 2003, even when controlling for the same time period (two weeks in August). In 2003, the times between eruptions were generally longer than in 1978 (median of 91 minutes versus 75 minutes). While about 25% of the intereruption times were less than an hour in 1978, only 2 of the 95 intereruption times in 2003 were less than an hour. That's bad news for the tourists, but the good news is that the geyser appears to be more predictable in 2003 than in 1978, because the variability in intereruption times is smaller as shown by the width of the boxplots (IQR of 11 minutes versus 23 minutes) and the standard deviations (8.46 minutes versus 12.97 minutes). Moreover, the histograms reveal that the intereruption times between distributions are *bimodal*, meaning that they have two peaks. In 1978, many intereruption times were around 50-60 minutes, and many were around 70-90 minutes, but relatively few were between 60-70 minutes. This gap and bimodality are much less pronounced but still present in 2003, another difference between the distributions. This bimodality is still present even for a larger number of observations during the year. You may have also noticed that the histogram became taller and "smoother" when we looked at a larger number of observations.

Discussion: While boxplots are very useful for comparing distributions, one limitation is that they conceal information about the shapes of the distributions. The histograms clearly reveal the bimodality of the distributions, which is completely hidden in the boxplots.

Investigation 2-4: Bumpiness, Variety, and Variability

Consider the following six histograms of (hypothetical) quiz scores in six classes (scores are integers ranging from 1 to 9):

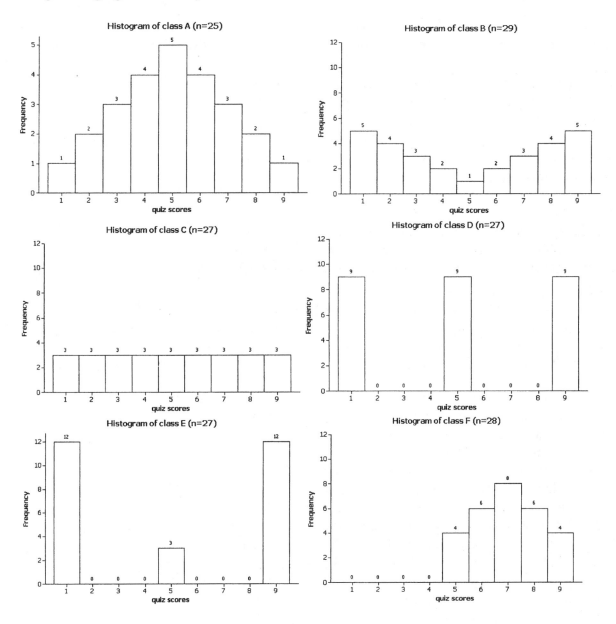

(a) Among classes A, B, and C, which do you think has the most variability in quiz scores? Which has the least? Or are they all the same? Explain your answers.

(b) Between classes C and D, which do you think has more variability in quiz scores? Explain.

(c) Between classes D and E, which do you think has more variability in quiz scores? Explain.

(d) How do you think class F's variability in quiz scores compares to the variability in the other classes? Explain.

(e) Determine the IQR of the quiz scores for each class by hand and record the results in the table below.

	Class A	Class B	Class C	Class D	Class E	Class F
Q_1						
Q_3						
IQR=Q_3-Q_1						

Based on these calculations, do you need to re-think any of your answers above? Explain.

(f) The following table displays the standard deviation of quiz scores for each class.

	Class A	Class B	Class C	Class D	Class E	Class F
Std. dev.	2.041	3.047	2.631	3.328	3.843	1.277

Based on these values, do you need to re-think any of your answers above? Explain.

Discussion: This activity addresses several misconceptions that many people have about the concept of variability. Many people confuse variability with "bumpiness." For example, the quiz scores in class A display much more bumpiness than those in class C, where heights of the bars are all equal. Nonetheless, class A's scores have less variability than class C's scores because more of them are clustered near the center. Keep in mind that variability deals with the spread along the horizontal axis, not the vertical axis.

Some people also confuse variability with "variety." The quiz scores in class C have many more different possible values (more "variety") than class D or E (just 3 different quiz scores in each of these). Nevertheless, class C's scores have less variability than class D's or class E's because more of them are clustered close to the center. Between classes D and E, they have the same number of different quiz scores, but most of the scores for class E are at the extreme, very few in the middle, leading to larger measures of variability. Keep in mind that variability deals with the distance from the center, not the number of distinct outcomes. Thus, a distribution like that of class F has very small variability since the values are much more tightly clustered around the mean. Note, that even though the value of the mean is larger, that doesn't necessarily lead to a higher measure of variability.

Def: A distribution whose histogram has constant height is often referred to as an *even* or *uniform* distribution. It is symmetric, but not all symmetric distribution has this evenness of uniformity.

Class C has a uniform distribution. It is often temping to refer to the shape of histograms A and F as "even" when you mean symmetric. But whenever you say "even," statisticians will picture class C instead of class A or F.

Practice:

2-3) Comparing Temperatures
The following table reports the average monthly temperature for San Francisco, CA and Raleigh, NC. Dotplots of these twelve temperatures for each city appear below.

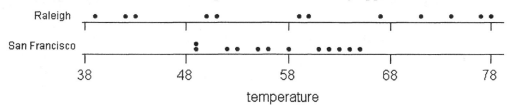

(a) Write a paragraph comparing the shape, center, and spread of these distributions. Support your conclusions with numerical calculations.
(b) Which is the more interesting comparison here, the centers of the distributions or the variability of the distributions? Explain.

2-4) Geyser Eruptions (cont.)

In the `oldfaithful.mtw` worksheet, recreate the 1978 histogram as in the Investigation.
(a) For the 1978 data, double click on the *x*-axis scale to bring up the "Edit Scale" box. Click the Binning tab and change the Interval Type from "Midpoint" to "Cutpoint." Describe how the histogram display changes.
(b) Pull up the Edit Scale box again and click the Binning tab again. In the Interval Definition box, select the "Number of Intervals" option. Change the number of intervals value from 14 to 6. Click OK. How did the histogram display change?
(c) Describe the distribution of intereruption times for 1978 based on the histogram in (b). Is the description the same as you made in the Investigation? Explain.

2-5) More Quiz Scores Reconsider the hypothetical quiz scores for classes A-F.
(a) For each class, calculate the range of the quiz scores.
(b) Is the range a helpful measure here in comparing the variability of these distributions? Explain.

2-6) Cloud Seeding (cont.)

Reconsider the *transformed* rainfall data from Investigation 2-1.
(a) Use Minitab to unstack the data so you can look at the groups individually:

```
MTB> unstack c3 c4 c5;    #assuming the transformed data are in C3
SUBC> subs c2.
MTB> name c4 'seeded' c5 'unseeded'
```

Produce dotplots or histograms for each column. Do you consider these distributions fairly symmetric?
(b) Recall that the mean and standard deviation for the seeded clouds was 5.134 inches and 1.60 inches. Use Minitab to determine the percentage of observations that fall within 1 standard deviation of the mean of the distribution:

```
MTB> let c6 = (c4>5.134-1.60 & c4<5.134+1.60)
MTB> tally c6
```

and convert the tally information into a percentage. Report this percentage.
(c) Repeat (b) for the unseeded clouds (mean 3.990 acre-feet, std dev 1.642 acre-feet). How do the two percentages compare?

Discussion: While the interquartile range has a nice interpretation in terms of measuring the width of the middle 50% of the data, we can apply a similar interpretation with the standard deviation *as long as the distribution is symmetric*. For symmetric data, approximately 68% of the observations should fall within 1 standard deviation of the mean of the distribution, i.e., in the interval (mean $\pm$ standard deviation).

Investigation 2-5: Body Temperatures

A study reported in the *Journal of the American Medical Association* (1992) called into question the traditional value of 98.6° as the typical healthy body temperature. The researchers cite problems with Carl Wunderlich's "axioms on clinical thermometry" and claim that the traditional value is out of date. Subjects were healthy men and women, aged 18-40 years, who were volunteers participating in Shigella vaccine trials at the University of Maryland Center for Vaccine Development, Baltimore. The researchers took the subjects' normal oral temperatures at several different times during three consecutive days using a digital thermometer. The data in temps.mtw (degrees Fahrenheit) were constructed by Shoemaker (1996) to match the histograms and summary statistics given in the article as closely as possible, but ensuring an equal number of men and women (65).

(a) Suppose you take your body temperature. For what temperature values would you be concerned about your health?

(b) Open temps.mtw and produce numerical and graphical summaries to compare the body temperatures for the men and women in this sample. Write a paragraph summarizing your comparison.

(c) Suppose someone registers a temperature of 98.6 degrees. Would you be concerned about this person's health based on these data? Explain. What if someone registers a temperature of 37 degrees Centigrade? If you are not sure, what other information would you want to know?

98.6°F:

37°C:

An observation of 98.6° is further from the average male temperature than from the average female temperatures (the females have a higher mean and median). Furthermore, a temperature of 98.6° is around the third quartile for the males but is below the third quartile for the females. Since the sample sizes have been made equal, we can directly compare the number of temperatures that fall below 98.6 degrees for each gender—20 men and 29 women. Thus, we might consider this temperature more worrisome for a male than a female. This situation arises because male temperatures are typically slightly lower and are a bit less variable than the female temperatures (SD = .699° vs. .743°). While 98.6° is above average for both genders, we would like a way of measuring "how much" above average that takes into account the variability in the data set.

Def: The *standard score* of an observation determines the number of standard deviations between the observation and the mean of the distribution:
$$z = \frac{\text{observation} - \text{mean}}{\text{standard deviation}}$$
This quantity is also referred to as the *z*-score. By converting to this *z*-score, we say we have *standardized* the observation.

(d) Calculate the *z*-score of a 98.6° temperature for a woman and for a man.

Female: Male:

(e) Based on your calculation in (d), 98.6° appears "further" from which mean? Explain.

Discussion: The main advantage of *z*-scores is that now the values are "on the same scale" and we can directly compare them. We see that 98.6° is .71 standard deviations from the male average, whereas it is only .28 standard deviations from the female average. This provides a measure of how far the observation of 98.6° is from the corresponding distribution means.

(f) Calculate the *z*-score of a 98° temperature for a woman and for a man. For which gender does this appear to be a more unusual temperature?

Female: Male:

More unusual?

(g) If a *z*-score is negative, what does this tell you about the relationship between the observation and the mean of the distribution?

Now suppose the temperatures had been reported on the Centigrade scale instead of the Fahrenheit scale.

(h) Use the `let` command to convert these results to degrees Centigrade, remembering that the conversion is: degrees Centigrade $= (5/9)$(degrees Fahrenheit $- 32$)
Determine the mean and standard deviation of these new values for the men and women.

	Mean ($^{\circ}$C)	Standard deviation ($^{\circ}$C)
Female		
Male		

(i) How do these mean and standard deviation values compare to those in the Fahrenheit scale?

(j) The temperature 98.6°F corresponds to what value on the Centigrade scale?

(k) Standardize this observation using the means and standard deviations you found in (h). How do they compare to the z-scores in (d)?

Female: Male:

Comparison:

If we apply a *linear* transformation to the data set (such as the conversion to Centigrade), the new mean is found through the same transformation of the original mean and the new standard deviation is scaled by the multiplicative factor but is not affected by the shift. If $y = a + bx$, then
$$\bar{y} = a + b\bar{x} \text{ and } SD(y) = bSD(x).$$
Thus values measured in either scale will have the same z-scores once standardized.

Discussion: A large advantage of z-scores is that they allow us to compare variables that may be recorded on completely different scales. For example, we could use z-scores to compare someone's temperature on the Fahrenheit scale to someone else's temperature on the Centigrade scale.

PROPERTIES OF Z-SCORES

(l) What is the value of the z-score at the mean of the distribution?

(m) If the distribution is symmetric, approximately what percentage of observations should have a z-score between -1 and 1? Explain.

Discussion: Since we must always consider the variability in a data set, the z-score will be the "ruler" by which we measure distances. An observation that may seem surprising in one data set may be judged differently in another data set depending on the center and spread of each distribution. The value of the z-score will be negative if an observation falls below the mean and will be positive if an observation falls above the mean. Practice problem 2-6 showed that with symmetric distributions, approximately 68% of the observations will fall within one standard deviation of the mean, or will have a z-score between -1 and 1. You will also see that approximately 95% of observations will fall within 2 standard deviations and approximately 99.7% will fall within 3 standard deviations for mound-shaped, symmetric distributions.

> The *empirical rule* says that, in a mound-shaped, symmetric distribution, approximately 68% of observations fall within 1 standard deviation of the mean, approximately 95% of observations fall within 2 standard deviations of the mean, and approximately 99.7% of observations fall within 3 standard deviations of the mean.

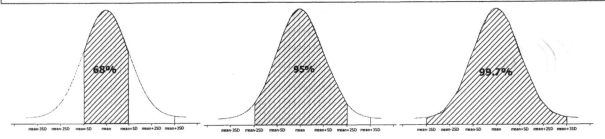

Thus, we will not consider an observation unusual until it is more than 2 standard deviations from the mean. In fact, another way to flag outliers is to see whether they are more than 2 standard deviations from the mean. Z-scores are most appropriate with symmetric distributions but can be used in other situations as well.

Study Conclusions: The body temperatures of men and women appear similar, though the female temperatures tend to be slightly higher and more variable. A temperature of 98.6° is above the mean for both distributions but falls within one standard deviation of the mean for each ($z = .708$ and $z=.277$ for men and women respectively). The z-score indicates that we would measure the "distance" between 98.6° and the mean to be larger for the men than for the women. We might hesitate to generalize this conclusion to a larger population because the subjects were volunteers in one study at the University of Maryland.

Practice:

2-7) SAT vs. ACT

Suppose the distribution of SAT scores is mound-shaped and symmetric with a mean of 1000 and a standard deviation of 200 and that the distribution of ACT scores is mound-shaped and symmetric with a mean of 28 and a standard deviation of 8. Suppose Tory scores a 1300 on the SATs and Jeff scores a 37 on the ACT.

(a) Provide a rough sketch, labeling the horizontal axis, of each distribution and indicating where the observed test score falls on the distribution.

(b) Which test taker had a higher score relative to the distribution of scores on that test? Explain. [*Hint*: Compare their *z*-scores.]

2-8) Body Temperatures (cont.)

(a) Determine the median and IQR of all 130 body temperatures on the Fahrenheit scale.

(b) Determine the median and IQR of all 130 body temperatures on the Centigrade scale.

(c) How do the values in (a) and (b) compare? Suggest a relationship between the new median and IQR values on the two scales.

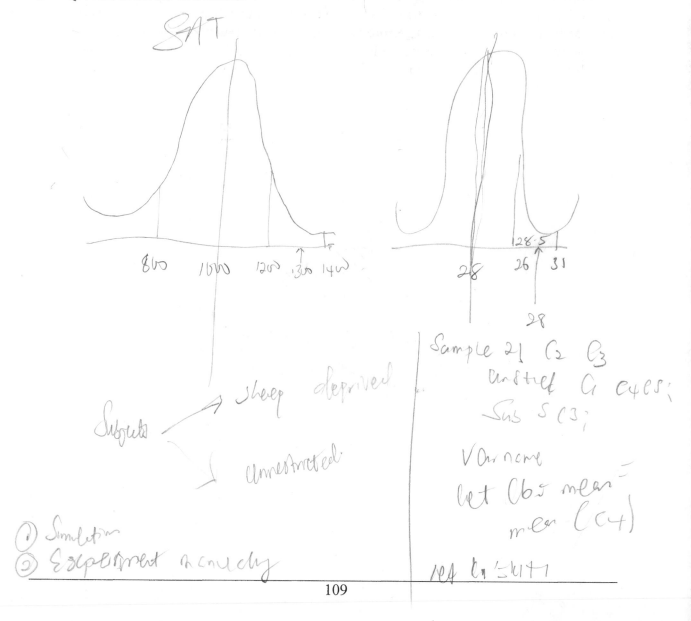

Investigation 2-6: The Fan Cost Index

The *Team Marketing Report* newsletter annually reports on the average cost of a family of four going to a Major League baseball game (www.teammarketing.com/fci.cfm). The "fan cost index" is a combination of costs of tickets, soda, beers, hot dogs, parking, caps, and a program. An obvious comparison to make is whether these costs differ between American League and National League teams. The 2003 fan cost values are in `fan03.mtw`. Open this worksheet.

(a) To calculate the average cost for a family of four to attend a game, the "fan cost index" (FCI) calculates the cost of 2 adult tickets, 2 kids tickets, parking, 2 programs, 2 caps, 2 small beers, 4 small sodas, and 4 hot dogs. To create a new column that calculates this value for each team, choose Calc > Calculator. Indicate that you want to store the result in variable fci. In the Expression box, enter the fan cost index, e.g., typing "2*" and then double clicking on "avg adult ticket" to include the two adult tickets, then type "+2*" and so on.

(b) Use Minitab to create parallel dotplots of the FCI values for each league. To identify unusual observations, change to "brush mode" by choosing Editor > Brush. Then choose Editor > Set ID Variables and enter C1 in the Variables box. Now place your cursor over any points you want to identify. Minitab will report the row number and the corresponding team. Identify, by team name, any outliers for these two distributions.

(c) Use Minitab to create boxplots of the FCI values for each league. Minitab uses the 1.5IQR criterion to flag outliers in the boxplot. Are any of the points that you identified in the dotplot above flagged as outliers?

(d) Use Minitab to produce summary statistics of the FCI values for the two leagues. Report the five number summary as well as the mean and standard deviation [*Habit:* remember to report the units of measurement as well].

(e) Based on these graphs and numerical summaries, describe and compare the distributions of the fan cost index values between the two leagues. [Remember to focus on center, spread, and shape.]

(f) If we compare the means of the FCI values, which league has a larger "center"? What if we compare medians? Why do you think the ordering reverses for these data? [*Hint*: Think about how the mean and median are calculated and about the shape of the distribution of these data.]

(g) Which league has the larger IQR of their FCI values? Which league has the larger standard deviation? Is the same measure larger in both leagues?

(h) How do you think the mean would change if Boston was removed from the data set? How do you think the median would change? The interquartile range? The standard deviation?

(i) In Minitab, click on the row number along the left edge of the data window to highlight all of row 5 and choose Edit > Cut Cells from the menu bar. Use Minitab to recalculate the descriptive statistics and re-examine the graphical summaries. Comment on whether and how they changed. Were your predictions in (i) correct? Are the comparisons between leagues now more consistent?

> **Def:** A numerical summary is considered *resistant* if it is not strongly affected by the presence of outliers in the data set.

(j) Between the mean and median, which is a resistant measure of center? Explain.

(k) Among the standard deviation, interquartile range, and range, which is a resistant measure of spread? Explain.

Comparing years

In the Data window, choose Edit > Paste Cells, choosing to re-insert Boston's 2003 FCI value above the active cell (248.44).

(l) Suppose we wanted to examine the change in the fan cost indices between the two years. Use Minitab to compute the price difference (2003-2002):

```
MTB> let c15 = c14-c12
```

and examine numerical and graphical summaries to compare the distributions of the price differences between the two years. Write a paragraph summarizing how the years compare.

> **Def:** The *percentage change* computes the difference and then adjusts for the magnitude of the current value by dividing: percentage change = (new value – old value)/old value × 100%.

(m) Use Minitab to compute the percentage change in FCI for each team:

```
MTB> let c16=c15/c12*100
```

Which team had the largest percentage change? Did this team have the largest 2003 FCI value? The largest change in FCI value? Suggest an explanation for the large percentage change for this particular team between 2002 and 2003.

Largest % change:

Largest 2003 FCI?   Largest change in FCI?

Possible explanation:

Other variable behavior

(n) Produce a dotplot of the program prices and a dotplot of the twill cap prices. We have been focusing on shape, center, spread, and outliers as the key descriptions of the distributions of a quantitative variable. Suggest another prominent feature of these distributions that stands out, other than shape, center, spread and outliers. Explain why it makes sense that cap prices and program prices at major league baseball games would have this feature.

(o) Identify at least one program price and two cap prices that do not follow the pattern in (n).

(p) Suggest an explanation for why these two teams have these unusual prices. [*Hint*: You may want to look back in the data file to identify these two teams and then discuss what they have in common that would lead to them having these unusual prices.]

(q) Produce a boxplot of the soda prices. Does Minitab identify any outliers in this distribution?

(r) Explain why it is not very useful to compare the soda prices given in column 6.

(s) Adjust for the size of the "small sodas" by creating a new variable: cost per ounce. Reproduce a boxplot. Identify by name any teams that are identified by outliers. In what ways are these team(s) unusual?

***Discussion*:** In the last few questions, you saw some different types of variability. For example, *granularity* occurs with the cap and program prices in that the prices tended to fall on the integer dollar amounts and not on the "in-between" values, leading to a distribution with a fixed amount of space between most of the values. You also considered choosing variables that are more appropriate measures to compare such as the cost per ounce rather than just the cost per soda since the sizes of "small sodas" varied across stadiums.

***Study Conclusions*:** In comparing the 2003 "fan cost index" values, we see that the behavior of the FCI values in the two leagues is pretty similar. Both distributions are fairly symmetric, though Boston is identified as an outlier according to the 1.5IQR criterion. The medians of the two distributions ($143.69 and $147.32) are similar. The spread is slightly larger for the national league (IQR = $37.79 versus $33.36) especially if Boston is removed from the data set (though we have no justification for taking this observation out of the data set). The medians and IQRs are slightly better to compare here (instead of the means and SDs) as they are resistant to the effect of outliers such as Boston. If we had compared the means, for example, the FCI of Boston was enough larger than the other American League values that it raises the entire average FCI value so that it is larger than the average National League FCI value, which is a little misleading. It is interesting to note that Boston is not an outlier when looking at the percentage change in prices (which takes into account the starting value of the FCI values instead of only examining the absolute change in prices) between 2002 and 2003. Some other interesting variable behavior includes the granularity in cap prices and program prices, presumably to make it easier on vendors to make change quickly, though this pattern is not followed by the Canadian teams. This is probably due to the conversion of Canadian dollars to American dollars. It's quite possible the amounts are integers in Canadian dollars. You also learned that not all "small sodas" are the same size and you can adjust for the size of the soda before making comparisons.

Practice:

2-9) "Me Cost Index"

Repeat your analysis of the fan cost index, but this time define a "me cost index" that represents the size and composition of a group with which you would like to attend a Major League Baseball game. Use the `let` command to create this new variable in, say, column 18, just as you calculated the FCI in C14, but changing the coefficients to match the purchases you believe your group would make. Name this column "mci".

(a) Explain and justify how you calculated your "me cost index."

(b) Where is it least expensive for your group to attend a game? What is the cost there? Where is it most expensive? What is the cost there? [*Hint*: Choose Data > Sort and double click on the "team" column and then on the "mci" column. Enter the "mci" column into the first "By column" box. Indicate that you want to store the sorted data in "column(s) of current worksheet" and enter C18 and C19 into the box. This should sort the teams in increasing order by cost, keeping the corresponding team name and "mci value" together.]

(c) Write a paragraph summarizing how the "me cost index" differs between the American and National Leagues. Remember to relate your comments to the context.

(d) Remove Boston from the data set again and discuss whether or not it is influential on your numerical summaries this time.

Investigation 2-7: House Prices

The median and mean both measure the center of a distribution of data, but you saw in the "cloud seeding" activity that they can produce quite different values. In this activity you will explore some different mathematical properties of the mean and median as measures of center.

Cal Poly students Peter Cerussi and Patrick Ziegler were interested in studying factors that are related to the price of a house. They gathered data from realestate.com on the listed prices of houses for sale in San Luis Obispo, CA on Nov. 20, 2003. The following dotplot displays the price distribution for eight of these houses (in thousands of dollars):

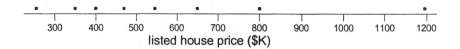

(a) Without performing any calculations, select one number that you think summarizes the center of this distribution.

(b) Do you think everyone in the class chose the same number? If not, how might we decide intuitively if one number does a "better" job than another at measuring the center?

A big issue in statistics involves deciding on a reasonable criterion by which to judge how well a prediction or estimate performs in the presence of variation. We don't expect to predict every value exactly, but can we minimize the amount of "error" in our predictions? Many reasonable criteria can be proposed, and choosing among them often amounts to deciding how serious the consequences are for one kind of an error versus another. Several types of criteria arise often enough to be worth studying.

One reasonable criterion involves looking at the differences or *deviations* between the data values (house prices) and your guess from part (a). The table below records the actual house prices displayed in the dotplot.

(c) Record the differences between these values and your guess (so that you get a negative difference for the houses that cost less than your guess for the center). Record these differences in the second row, and then calculate the sum of these differences and record it also:

Price	255	349	399	469	545	649	799	1195	Sum
Difference									

(d) If your guess really does a good job of measuring the center of the house price distribution, what would be a "good" value for this sum?

(e) Determine what value one's guess would have to be in order for this sum to equal zero. [*Hint*: One approach is to see how much above/below zero your sum is, and figure out how much you need to lower/raise your guess to make the sum equal zero instead.]

Let m represent a measure of the center of the distribution of house prices. We can express this sum of differences as a function of m:

$$\text{SumDiff}(m) = (255\text{-}m) + (349\text{-}m) + (399\text{-}m) + (469\text{-}m) + (545\text{-}m) + (649\text{-}m) + (799\text{-}m) + (1195\text{-}m).$$

(f) Simplify this expression algebraically, set SumDiff(m) equal to zero, and solve for m. Does this confirm your answer to (e)?

(g) Calculate the mean and median of these house prices. Does one of these equal your answers to (e) and (f)? If so, which one?

Now consider the general case of any data set with n observations, with x_i denoting the ith data value. The function SumDiff(m) can then be expressed as: $\text{SumDiff}(m) = \sum_{i=1}^{n}(x_i - m)$.

(h) Simplify the expression for this function SumDiff(m), set it equal to zero, and solve for m in terms of the x_i's.

Discussion: You should have found that the unique value that forces the sum of these differences to equal zero is the *mean*. Thus, one way to interpret the mean is that it is the value that "balances out" the differences on either side of it: the sum of the positive differences equals the sum of the negative differences, which makes the overall sum of the differences equal zero. For this reason the mean is often referred to as the balance point of the distribution.

Another reasonable criterion by which to judge your guess is to consider the sum of the *absolute* deviations between the data values and your guess from part (a).

(i) Calculate these absolute values and record them in the third row of the table below. Then calculate their sum and record it also:

Price	255	349	399	469	545	649	799	1195	Sum
Absolute deviation									

(j) What is the smallest sum of absolute deviations in your class? Report both the guess and this sum.

To search for a possibly even better guess according to this sum of absolute deviations criterion, let m represent a generic guess. Consider a function called SAD (sum of absolute deviations) defined just as you calculated for your guess above:

$$SAD(m) = |255 - m| + |349 - m| + |399 - m| + |469 - m| + |545 - m| + |649 - m| + |799 - m| + |1195 - m|.$$

(k) Open the Excel spreadsheet document "HousePrices.xls." Notice that the eight house prices are in column A, while column B contains values of m from 300 to 800. Click on cell C2, and notice that it contains a formula for $SAD(m)$, which refers back to the data values in column A and to the value of m from B2 (namely, $m=300$ there). Use Excel's "fill down" feature to calculate the values of $SAD(m)$ for the remainder of column C. [*Hint*: You can do this in one of several ways: With the C2 cell selected, double click on the box in the lower right corner, or pull the right corner of the highlighted C2 cell down to the end of the column, or highlight C2 and the cells to be filled and choose Edit> Fill> Down.] Confirm that the SAD value for your guess agrees with what you calculated above.

(l) Highlight columns B and C, holding the ctrl key down to select both at once. Use Excel's "chart wizard" to construct a graph of this function, selecting the "XY (Scatter)" type of graph, with data points connected by smooth lines (2nd option). [*Hint*: After you create the graph, you may want to change the x-min and x-max values by double clicking on the horizontal axis and choosing the scale tab.] Reproduce a rough sketch of this graph below, and comment on the shape of this function. Does it seem to follow a familiar form, such as linear or parabolic or exponential? Explain.

(m) Does there seem to be a unique value of m that minimizes this SAD(m) function? If so, identify it. If not, describe all values of m that minimize the function. [*Hint*: Pointing your mouse over a point displays its coordinates.]

(n) Look back at the dotplot and at the values of the actual house prices. What do you notice about the values of m for which the SAD function is minimized?

(o) Suppose that the most expensive house had been listed at 2195 thousand dollars rather than 1195. Make this change in the spreadsheet (cell A9), and note that the graph of the SAD function is updated automatically. [*Hint*: You may want to change the scale on the vertical axis by double-clicking on the axis and selecting the "scale" tab.] Now for what value(s) of m is SAD minimized, and what is that minimum value of SAD? Has either of these numbers changed?

(p) Change the most expensive house's value back to 1195, and now change the fourth house's price from 469 to 529 thousand dollars (cell A5), and re-evaluate SAD. Now what has changed, and what has not?

(q) Based on these findings, make a conjecture as to how one can determine from a generic set of data the value that minimizes the sum of absolute deviations from the data values.

(r) Consider the "mean absolute deviation" (MAD) criterion, which divides the SAD value by the number of observations. How would the MAD(m) function compare to the SAD(m) function? Would its minimum be achieved at the same value(s) of m? Explain.

Discussion: You should have found that the SAD function is *piecewise linear*. In other words, the function itself is not linear, but its component pieces are linear. In this case the function changes its slope precisely at the data values. The SAD function (and MAD function) are minimized at the *median* of the data and also at any value between (and including) the two middle values in the dataset (the 4th and 5th values in this case). It is a convenient convention to define the median as the average of those two middle values.

Another criterion is to find the value of *m* that minimizes the *sum of the squared deviations*.
(s) Square the differences from the table above and determine the sum for your guess.

Price	255	349	399	469	545	649	799	1195	Sum
Squared Deviation									

(t) Who has the smallest sum of squared deviations in your class? Is this the same person as before?

(u) Make sure that the data values in column A are back to their original values. Then click on cell D2 in the Excel spreadsheet, and notice that it gives the formula for the SSD (sum of squared deviations) function. Fill this formula down the column and use the chart wizard to create a second graph to investigate the behavior of this function. Is this function a polynomial? What is its shape? At what value of *m* is the function minimized?

(v) Increase the most expensive house's price by 800 (thousand dollars), and investigate the effect on the SSD function. Has the value of *m* that minimizes SSD changed? By how much? How does this compare to the effect on the SAD function in part (o)?

(w) Make a conjecture for determining the value *m* that minimizes SSD for a given set of data values.

(x) Use calculus to determine the value of *m* that minimizes SSD for an arbitrary set of data.

[*Hint*: Consider the function: $SSD(m) = \sum_{i=1}^{n}(x_i - m)^2$. Differentiate this function with respect to *m*, set the derivative equal to 0, and solve for *m*. Then find the second derivative and evaluate it at this value to verify that you have indeed found a minimum.]

Def: A common minimization criterion in statistics is the principle of *least squares*: minimize the squared deviations of data values from a prediction.

With one quantitative variable, the least squares estimator of the center of a distribution is the *mean* of the data values.

(y) When you changed the maximum value in the dataset in (v), which function changed more- SAD(m) or SSD(m)? Which measure of center changed more- the median or mean?

You should have noticed that the value of m for which the SSD function is minimized (the mean of the data set) shifted dramatically to the right as the maximum increased, but the value of m for which SAD is minimized (including the median of the data set) did not change at all.

This further illustrates that the mean is not resistant to outliers but the median is. Thus, if a data set has extreme outliers, then the median may be a more appropriate measure of center to report.

(z) Provide an intuitive explanation for why the mean is not resistant but the median is. That is, how would you explain this to someone who didn't want to deal with the mathematical formulas?

(aa) If you were looking to buy a house in this area, would you be more interested in the mean price or the median price? Which minimization criterion (or perhaps suggest another one) would be most important to you? Explain.

Practice:

2-10) House Prices (cont.) Add a ninth house price to the dataset: 419 thousand dollars.
(a) Use Excel to evaluate the SAD function for this new data set. Does the function now have a unique minimum? If so, at what value does that minimum occur?
(b) Based on your finding in (a), make a general conjecture as to the value that minimizes the SAD function where there is an odd number of observations.
(c) Use Excel to evaluate the SSD function for this new data set. At what value does the minimum occur? Is this consistent with your finding about where SSD was minimized with eight observations? Explain.

2-11) House Prices (cont.)
You will now consider other criteria based on the absolute deviations between the data values and your guess. Even if you keep absolute deviations as your basis for a minimization criterion, you can consider functions other than the sum. For example, if you want to be sure that you are never too far off, you might want to minimize the *maximum* of those absolute deviations:
$$\text{MAXAD}(m) = \max\{|255-m|,|349-m|,|399-m|,|469-m|,|545-m|,|649-m|,|799-m|,|1195-m|\}$$
(a) Use the Excel file "HousePrices.xls" to investigate the behavior of this MAXAD function. Return the data values (house prices) in column A to their original values, and click on cell E2. Notice that this cell contains a formula for evaluating the MAXAD function. Use the "fill down" feature to evaluate this function for the rest of the m values. Then use Excel to draw a graph of the MAXAD function. Describe its behavior, and comment on whether it has a unique minimum value. Identify where the minimum occurs and what that minimum value is.
(b) Change the maximum house price from 1195 to 895 thousand dollars. Comment on the impact of this change on the MAXAD function and especially on where it is minimized.
(c) Change the fourth house's price from 469 to 529 thousand dollars, and re-evaluate the MAXAD function. Now what has changed, and what has not?
(d) Now change the cheapest house's price from 255 to 305 thousand dollars, and re-evaluate the MAXAD function. Now what has changed, and what has not?
(e) Based on this analysis, make a conjecture for determining the value that will minimize the maximum of absolute deviations from the data values.

2-12) Create an Example
(a) Create a hypothetical example of ten exam scores (say, between 0 and 100 with repeats allowed) such that 90% of the scores are above the mean
(b) Repeat (a) for the condition that the mean is roughly 40 points less than the median.
(c) Repeat (a) for the condition that the IQR equals zero and the mean is more than twice the median.

Investigation 2-10: Lifetimes of Notables

Do scientists tend to live longer than writers? To investigate this conjecture we gathered data on the lifetimes of 20 writers and 20 scientists listed in the *World Almanac and Book of Facts*. The data are displayed in the following *side-by-side stemplot*. To interpret this display, each lifetime has been broken into a "stem" (tens digit) and "leaf" (ones digit). The stems are between the vertical lines in the middle of the display, and the leaves extend in order from the inside out, to the right for scientists and to the left for writers. Note the min and max values for the writers are 29 years and 90 years, respectively.

```
     Writers (n=20)        Scientists (n=20)
              9 | 2 |
              5 | 3 |
              3 | 4 | 8
              9 | 5 | 0389
        76622100 | 6 | 66
             751 | 7 | 0357789
            9530 | 8 | 679           Leaf unit = 1 year
               0 | 9 | 004
```

(a) Calculate the five-number summary of the lifetimes for each group (note that the stemplot automatically orders the observations):

	Minimum	Lower quartile	Median	Upper quartile	Maximum
Writers					
Scientists					

(b) Write a few sentences comparing and contrasting the distributions of lifetimes between these two groups. Be sure to address the question that motivated looking at these data in the first place: do scientists tend to live longer than writers? Also comment on other aspects (such as spread and shape) of the distributions.

(c) Do these data come from a controlled experiment or an observational study? Explain.

(d) Open the Minitab worksheet `lifetimes.mtw`. Note that the stacked data appear in C1 and the occupation categories in C2. Determine the difference in the mean lifetime of writers and scientists in this study and the difference in the median lifetimes.

(e) Modify your "randomize" macro so that it randomly assigns the occupation categories to the 40 lifetimes and so can be used to assess the statistical significance of the difference in mean lifetimes between these two groups. Then run the macro 1000 times. Comment on the distribution of differences in group means, and report the approximate p-value.

(f) Based on your simulation analysis, would you say that the difference in mean lifetimes is statistically significant? Explain.

(g) Even if the p-value were smaller, say less than .0001, would you conclude that the occupation is causing the difference in mean lifetimes? Would you advise a young writer that changing his or her profession to science would tend to cause a longer lifetime? Explain.

Study Conclusions: This example should remind you that cause-and-effect conclusions can not be drawn from observational studies. The stemplot reveals that scientists do tend to live longer than writers, and the difference in median lifetimes is 10 years (76 for scientists, 66 for writers). Both distributions are roughly symmetric, perhaps a bit skewed to the left. The lifetimes vary more for the writers in that they range from the 20's through 90, as opposed to scientists ranging from the 40's through the 90's, but on the other hand the writers' lifetimes have a strong concentration in the 60's. Neither group has any obvious outliers.

The simulation reveals that the approximate p-value for comparing the group means is about .07. This suggests that if there was no difference between the groups, it is unlikely but not terribly so for such a large difference to occur by chance. But even if the p-value were infinitesimally small, we still could not attribute the longer lifetimes to the choice of occupation, because this observational study does not control for confounding variables. One possible explanation for the observed tendency is that scientists require more formal training in order to succeed than writers, so if someone dies young but famous, they are more likely to have achieved their fame as a writer than as a scientist.

Practice:

2-13) Sleep Deprivation (cont.)
(a) Recall the difference in group medians between these two groups.
(b) Change one line of your Minitab macro from Investigation 2-8 so that it analyzes difference in group *medians* rather than differences in group means. Apply your macro to the original data from the sleep deprivation study, using at least 1000 repetitions. Describe the distribution of differences in group medians, and report the empirical p-value.
(b) Comment on how the randomization distribution of the differences in medians differs from your earlier analysis of means.
(c) What conclusion would you draw about the statistical significance of the observed difference in group medians from this analysis? Explain.
(d) Now apply the macro for analyzing differences in medians to the hypothetical data from Investigation 2-9. Again report the approximate p-value, and compare this analysis to your earlier analysis of means.
(e) Does your earlier conclusion that the difference is much less significant for the hypothetical data than for the real data hold for analyzing medians as well? Explain.

2-14) Cloud Seeding (cont.)
Recall the cloud seeding study of Investigation 2-1. The data are stored in
`CloudSeeding.mtw`.
(a) Simulate a randomization distribution to assess whether the difference in the *mean* rainfall amounts is statistically significant.
(b) Simulate a randomization distribution to assess whether the difference in the *median* rainfall amounts is statistically significant.
(c) Write a paragraph summarizing your results. In particular, would be it valid to attribute these differences to the imposition of the cloud seeding? Explain.

SUMMARY

In this chapter, you have again focused on comparing the outcomes for different groups. With quantitative data, the interesting comparisons include shape, center, and spread. The appropriate numerical summaries of center include the mean and median. You saw the median is more resistant than the mean to outliers and skewness. The appropriate numerical summaries of spread include interquartile range (IQR) and standard deviation, with the IQR also more resistant to outliers and skewness. Make sure you keep in mind that when examining the variability of a data set, we are interested in the spread of the values that the variable takes (along the horizontal axis). In summarizing data, the five number summary can be very descriptive. In this chapter, you were also introduced to boxplots (a picture of the five-number summary), histograms, and stemplots. You should examine these graphs to help you decide which numerical summaries are most appropriate. For example, both IQR and standard deviation measure spread (the IQR tells us the width of the middle 50% of the data and the standard deviation tells us about the width of the middle 68% for mound-shaped symmetric distributions). While the IQR is always applicable, standard deviation should only be used for mound-shaped, symmetric distributions. Another use of the standard deviation is to help us calculate z-scores as a measure of distance between an observation and the mean of a distribution. The role of z-scores will be greatly expanded in later chapters.

After exploring the data sets, it may be appropriate to ask whether the differences observed could have arisen *by chance*. In other words, is the observed difference big enough to convince us that it arose from a genuine difference in the groups instead of from the randomization inherent in designing the experiment? In this chapter, we focused on the difference in means and the difference in medians through a randomization test. Just as in chapter 1, we used simulation to approximate how often we would obtain a difference in means at least as extreme as observed just by chance, and then we used enumeration to calculate the p-value exactly. In chapter 4, you will learn other techniques for calculating such p-values. Keep in mind that the logic of statistical significance hasn't changed—we assume there is no effect and determine how often we would get a result at least as extreme as what was observed. You saw in Investigation 2-9 that this chance is strongly affected by the amount of variability in the groups. It is very important when reporting on your statistical study that you also mention the sample sizes. You were also reminded in the last activity that no matter how small a p-value is, we cannot draw cause-and-effect conclusions unless the data came from a properly designed randomized experiment.

TECHNOLOGY SUMMARY:

- You learned many Minitab tools for creating numerical and graphical summaries including boxplots, dotplots, and histograms (all under the Graph menu) and displaying descriptive statistics (Stat > Basic Statistics > Display Descriptive Statistics). In particular, you learned how put these distributions on the same scale.

- You also practiced "stacking" and "unstacking" columns in Minitab.

- You used the "log" command to transform a variable.

- You used a Minitab macro to simulate a randomization distribution by randomizing the column containing the explanatory variable.

- You used some of the automatic updating features in Excel to examine the behavior of different functions.

CHAPTER 3: SAMPLING FROM POPULATIONS

The previous two chapters have been about comparing groups. You learned about numerical and graphical summaries of categorical (chapter 1) and quantitative (chapter 2) variables, as well as how to decide whether the difference between two groups is statistically significant. One important caution in interpreting those results was that you often dealt with volunteers or observational data so that you were not be able to generalize beyond that particular group of observational units. In this chapter, you will focus on how to select samples from a larger population so that you may generalize the sample results back to that population.

SECTION 3-1: SELECTING SAMPLES FROM POPULATIONS I

Investigation 3-1: Sampling Words

(a) Circle 10 representative words in the following passage.

Four score and seven years ago, our fathers brought forth upon this continent a new nation, conceived in liberty, and dedicated to the proposition that all men are created equal.

Now we are engaged in a great civil war, testing whether that nation, or any nation so conceived and so dedicated, can long endure. We are met on a great battlefield of that war.

We have come to dedicate a portion of that field as a final resting place for those who here gave their lives that that nation might live. It is altogether fitting and proper that we should do this.

But, in a larger sense, we cannot dedicate, we cannot consecrate, we cannot hallow this ground. The brave men, living and dead, who struggled here have consecrated it, far above our poor power to add or detract. The world will little note, nor long remember, what we say here, but it can never forget what they did here.

It is for us the living, rather, to be dedicated here to the unfinished work which they who fought here have thus far so nobly advanced. It is rather for us to be here dedicated to the great task remaining before us, that from these honored dead we take increased devotion to that cause for which they gave the last full measure of devotion, that we here highly resolve that these dead shall not have died in vain, that this nation, under God, shall have a new birth of freedom, and that government of the people, by the people, for the people, shall not perish from the earth.

The authorship of literary works is often a topic for debate. Were some of the works attributed to Shakespeare actually written by Bacon or Marlowe? Which of the anonymously published *Federalist Papers* were written by Hamilton, which by Madison, which by Jay? Who were the authors of the writings contained in the Bible? The field of "literary computing" examines ways of numerically analyzing authors' works, looking at variables such as sentence length and rates of occurrence of specific words.

The above passage is of course Abraham Lincoln's Gettysburg Address, given November 19, 1863 on the battlefield near Gettysburg, PA. In characterizing this passage, we would ideally examine every word. However, often it is much more convenient and even more efficient to only examine a subset of words. In this case you will examine data for just 10 of the words. We are considering this passage to be a *population* of 268 words, and the 10 words you selected are therefore a *sample* from this population.

In most statistical studies, we do not have access to the entire population and can only consider data for a sample from that population. For example in Chapter 1, we only examined 116 current popcorn plant workers - not all previous workers, and not all plants; in Chapter 2, the sleep researchers only performed their experiment on 21 students rather than all students at their university. Our ultimate goal is to make conclusions about a larger population or overall process, even when we only have access to a sample from the population.

(b) Consider the following variables:
- length of word (number of letters)   Type: *quantitative*
- whether or not word contains more than 4 letters   Type: *Categorical*

Identify each variable as quantitative or categorical.

Record the data from your sample for the above variables:

	1	2	3	4	5	6	7	8	9	10
Word										
Length										
>4 characters?										

Ideally, we want our sample to be *representative* of the population, that is, having the same characteristics.

(c) Do you think the words you selected are representative of the population of 268 words in this passage? Explain.

(d) Construct a dotplot of the distribution of the word lengths in your sample. Also calculate the sample mean length and describe the characteristics of this distribution. [Remember to label your plot and to relate your comments to the context.] What are the observational units in this graph?

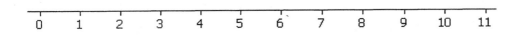

To display the distribution of a categorical variable for just one sample, we can construct a *bar graph*. This graph has a separate bar for each category, with heights corresponding to the proportion of observational units in that category.

(e) Construct a bar graph for the categorical variable (whether or not the word is "long" – more than four characters) for your sample and describe its distribution. [Remember to label your plot and to relate your comments to the context.] What are the observational units in this graph?

Def: A *parameter* is a numerical characteristic of the population. A *statistic* is a numerical characteristic of the sample. We usually denote population parameters with Greek letters, e.g., π or μ for a proportion or mean, respectively. We denote the statistics for a sample proportion and a sample mean by $\hat{p}$ and $\bar{x}$, respectively.

(f) Is the average length you calculated in (d) a parameter or a statistic? Explain. What symbol do we use to denote this value?

(g) Is the proportion of long words you calculated in (e) a parameter or a statistic? Explain. What symbol do we use to denote this value?

(h) The average length of all 268 words in this population is 4.29 letters. Is this number a parameter or a statistic? What symbol do we use to denote this value?

(i) There are 99 "long" words in this population of 268 words. What proportion of the words in the population are "long"? Is this number a parameter or a statistic? What symbol do we use to denote this value?

(j) Did everyone in your class obtain the same value for the sample mean? The same sample proportion?

(k) Construct a dotplot or histogram combining the average length of words in your sample with those of your classmates. Be sure to label the horizontal axis, and also indicate where the population mean falls. What are the observational units in this graph? Describe the distribution of these sample means, particularly with regard to where the population average falls.

(l) Construct a dotplot combining the proportion of long words in your sample with those of your classmates. Be sure to label the horizontal axis, and also indicate where the population proportion falls. What are the observational units in this graph? Describe the distribution of these sample proportions, particularly with regard to where the population proportion falls.

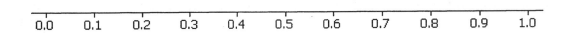

> You have witnessed the fundamental principle of *sampling variability*: values of sample statistics *vary* when one repeatedly takes random samples from a population. A key point in analyzing these results is that now we are treating the samples as the observational units and the sample statistics as the variable of interest (so the first graph label should be something like "sample means" and the second label should be something like "sample proportions"). Many statistical methods are based on describing the pattern of variation of the sample results. Looking at the resulting pattern of variation in sample statistics is also one way to determine if a sampling method is reasonable.

(m) Was your sample average word length $\bar{x}$ above or below the population mean μ? How many and what proportion of students in your class found a sample mean word length that exceeded the population mean?

(n) Was your sample proportion of long words $\hat{p}$ above or below the population proportion π? How many and what proportion of students in your class found a sample proportion of long words that exceeded the population proportion?

(o) Based on your answers to (k) - (n), would you say that this sampling method (asking people to circle 10 words) is likely to produce a sample that is truly representative of the population with respect to length of the words? Explain.

Def: When characteristics of the resulting samples are systematically different than characteristics of the population, we say that the sampling method is *biased*. When the distribution of the sample statistics is centered at the value of the population parameter, the sampling method is said to be *unbiased*.

For example, we suspect that your class repeatedly and consistently overestimated the average length of a word and the proportion of long words. Not everyone has to overestimate, but if there is a *tendency* to err in the same direction time and time again, then the sampling method is biased. In other words, sampling bias is evident if we repeatedly draw samples from the population and the distribution of the sample statistics is not centered at the population parameter of interest. Note that bias is a property of a sampling *method*, not of a single sample. Studies have shown that human judgment is not a good basis for selecting representative samples, so we will rely on other techniques to do the sampling for us.

(p) Consider another sampling method: you close your eyes and point at the passage on p. 3-1 and select whatever word your pen lands on. Would this sampling method be biased? If so, in which direction? Explain.

Discussion: Even though the method in (p) sounds "random," it is also going to be biased because longer words take up more space on the page and so will be more likely to be selected than shorter words. Once again, there will be a tendency to oversample the longer words and thus to repeatedly overestimate the average length and the proportion of long words.

(q) Suggest a better method for selecting a sample of 10 words from this population.

Def: A *simple random sample* gives every observational unit in the population the same chance of being selected. In fact, it gives every sample of size *n* the same chance of being selected. So any set of 10 words is equally likely to end up as our sample.

"Low-tech" methods for obtaining a simple random sample from a population include using a *random number table* (see Table I) or a calculator   A random number table is constructed so that each position is equally likely to be filled by any digit from 0 to 9, and the digit in one position is unaffected by the digit in any other positions. The first step is to obtain a list of every member of your population (this list is called a *sampling frame*). Then, give each observational unit on the list an ID number, and use the random number table to select *n* unique ID numbers. If an ID number occurs more than once, continue reading on to the next ID number. If a number from the random number table does not correspond to a member of the population, continue reading on to the next ID number.

The following is a sampling frame for the Gettysburg address, with each word in the population numbered.

1 Four	35 in	69 dedicate	103 But,	137 add	171 here	205 these	239 that
2 score	36 a	70 a	104 in	138 or	172 to	206 honored	240 this
3 and	37 great	71 portion	105 a	139 detract.	173 the	207 dead	241 nation,
4 seven	38 civil	72 of	106 larger	140 The	174 unfinished	208 we	242 under
5 years	39 war,	73 that	107 sense,	141 world	175 work	209 take	243 God,
6 ago,	40 testing	74 field	108 we	142 will	176 which	210 increased	244 shall
7 our	41 whether	75 as	109 cannot	143 little	177 they	211 devotion	245 have
8 fathers	42 that	76 a	110 dedicate,	144 note,	178 who	212 to	246 a
9 brought	43 nation,	77 final	111 we	145 nor	179 fought	213 that	247 new
10 forth	44 or	78 resting	112 cannot	146 long	180 here	214 cause	248 birth
11 upon	45 any	79 place	113 consecrate,	147 remember,	181 have	215 for	249 of
12 this	46 nation	80 for	114 we	148 what	182 thus	216 which	250 freedom,
13 continent	47 so	81 those	115 cannot	149 we	183 far	217 they	251 and
14 a	48 conceived	82 who	116 hallow	150 say	184 so	218 gave	252 that
15 new	49 and	83 here	117 this	151 here,	185 nobly	219 the	253 government
16 nation:	50 so	84 their	118 ground.	152 but	186 advanced.	220 last	254 of
17 conceived	51 dedicated,	85 their	119 The	153 it	187 It	221 full	255 the
18 in	52 can	86 lives	120 brave	154 can	188 is	222 measure	256 people,
19 liberty,	53 long	87 that	121 men,	155 never	189 rather	223 of	257 by
20 and	54 endure.	88 that	122 living	156 forget	190 for	224 devotion,	258 the
21 dedicated	55 We	89 nation	123 and	157 what	191 us	225 that	259 people,
22 to	56 are	90 might	124 dead,	158 they	192 to	226 we	260 for
23 the	57 met	91 live.	125 who	159 did	193 be	227 here	261 the
24 proposition	58 on	92 It	126 struggled	160 here.	194 here	228 highly	262 people,
25 that	59 a	93 is	127 here	161 It	195 dedicated	229 resolve	263 shall
26 all	60 great	94 altogether	128 have	162 is	196 to	230 that	264 not
27 men	61 battlefield	95 fitting	129 consecrated	163 for	197 the	231 these	265 perish
28 are	62 of	96 and	130 it,	164 us	198 great	232 dead	266 from
29 created	63 that	97 proper	131 far	165 the	199 task	233 shall	267 the
30 equal.	64 war.	98 that	132 above	166 living,	200 remaining	234 not	268 earth.
31 Now	65 We	99 we	133 our	167 rather,	201 before	235 have	
32 we	66 have	100 should	134 poor	168 to	202 us,	236 died	
33 are	67 come	101 do	135 power	169 be	203 that	237 in	
34 engaged	68 to	102 this.	136 to	170 dedicated	204 from	238 vain,	

(r) What is the largest ID number? How many digits are in that ID number?

(s) Use the random number table to select a simple random sample of 5 words from this population. Do this by choosing a row and finding 5 different three-digit ID numbers, throwing away any numbers above 268 or repeated three-digit numbers, continuing on to the next row as needed. Write down resulting ID numbers, the corresponding words, the length of each word, and whether or not it is long.

	1	2	3	4	5
ID number					
Word					
Length					
>4 characters?					

(t) Calculate the average word length and the proportion of "long" words in this sample of 5 words. Again combine your results with your classmates to produce a dotplot of the sample means and a dotplot on the sample proportions. Comment on each distribution. How do these distributions compare to the ones in (k) and (l)?

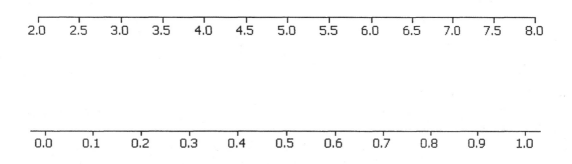

(u) When taking random samples, did everyone obtain a sample mean equal to the population mean? Did everyone in your class obtain the same sample mean in the random samples? In what way would you say these random samples produce "better" sample results?

(v) What proportion of your class obtained a sample average that was larger than the population mean? What proportion of your class obtained a sample proportion that was larger than the population proportion? Does this sampling method appear to be biased? Explain.

$\frac{15}{100}$ 0.15

(w) Is it plausible that the sampling method of randomly selecting just 5 words will have less bias than the sampling method of selecting 10 words you used did in (a)? Explain.

Summary: Even with randomly drawn samples, sampling variability still exists (*random sampling errors*). While not every random sample produces the same characteristics as the population, the goal is for the sample statistics to not consistently overestimate the value of the parameter or consistently underestimate the value of the parameter. You should have found that when we *randomly* select the samples, you will not have the same pattern of consistently overestimating the value of the population parameter. A *simple random sample* eliminates bias by giving every sample of size *n* the same probability $1/C(N, n)$ of being selected. When the distribution of the sample statistics is centered at the value of the population parameter, the sampling method is said to be *unbiased*. When the sampling method is unbiased, we expect the individual samples to have characteristics similar to those of the population. Thus, when we select just one sample, we will feel comfortable generalizing the results from that one sample to the larger population. If the sampling method is biased, we can make no claims about the population parameter.

Furthermore, a second virtue of random samples is that we will see that sampling variability follows a predictable, long-term pattern. In particular, with random samples we will be able to estimate how far the sample statistic is likely to fall from the population parameter, that is, the size of the *random sampling error,* and what factors affect the size of this random sampling error.

In this example, we knew the value of the population parameter, but that is not usually the case. Thus, it is very important to determine whether or not the sample was selected at random before you believe that the sample results are representative of the population.

Keep in mind that *random sampling*, as you have done here, and *randomization*, as you studied in Chapters 1 and 2, are different uses of randomness. Now we are focusing on selecting a subset from a larger population, and making inferences back to that population instead of trying to compare two groups as you did earlier.

Notice that the ideas of a parameter and a statistic are closely related; one describes the population and one describes the sample. For example, the average length of words in the population is a parameter, but the average length of words in the sample is a statistic. A crucial observation is that the value of a statistic will vary from sample to sample (we are now thinking of the statistic as a variable but with the samples as the observational units!). But as we saw, when we take random samples, this variation follows a very predictable pattern. Keep in mind that although the value of a statistic varies from sample to sample, a parameter does not vary.

Practice:

3-1) *Literary Digest*

In 1936, *Literary Digest* magazine conducted the most extensive (to that date) public opinion poll in history. They mailed out questionnaires to over 10 million people whose names and addresses they obtained from telephone books and vehicle registration lists. More than 2.4 million responded, with 57% indicating that they would vote for Republican Alf Landon in the upcoming presidential election.

(a) Identify the population of interest, the sample, and the sampling frame in this study. Also define the parameter and statistic.

(b) Incumbent Democrat Franklin Roosevelt won the election, carrying 63% of the popular vote. Give two explanations why the *Literary Digest* prediction was so much in error. In particular, talk about the direction of the bias – why was this sampling method vulnerable to producing an overestimate of the parameter?

3-2) Sampling Senators

Suppose we want a sample of 5 members of the U.S. Senate and decide to ask you to name 5 senators. We then determine how many years these senators had served in the Senate and calculate the average service time.

(a) Identify the population, sample, parameter, and statistic in this study.

(b) Do you think this sampling method will be biased? If so, do you think it will tend to overestimate or underestimate the parameter of interest? Explain.

Investigation 3-2: Comparison Shopping

Suppose you wanted to compare the prices at grocery stores in your area. You have obtained an inventory list from Store L and plan to select about 30 products from this store.

(a) Identify the observational units, the sample, the population, and the sampling frame for this study.

(b) Explain how you could obtain a simple random sample of 30 products from Store L.

(c) Imagine you now have to find the prices of the items for this random sample, what is a potential inconvenience in obtaining these prices?

(d) Consider another sampling method: the inventory list is broken into 30 pages with 35 items on each page. We could randomly select a number between 01 and 35 and then take that numbered item from each page. Would the selection of the sample proceed any more quickly than with the plan in (b)? Would gathering prices for these items be any more convenient than the plan in (b)? Explain.

Def: The sampling scheme described in (d) is a *systematic sample*. If you are using a sampling frame, a systematic sample divides the list into equally sized segments (say m items in each segment) and then takes the kth item from each segment where k is selected randomly from the digits 1 to m. This sampling method is often as efficient as a simple random sample (and more convenient), but you must be cautious of any periodicity or cyclic patterns in the data.

(e) Consider a third sampling technique. Keep in mind that the aisles are numbered, that there are a fixed number of shelves in each aisle, and that the aisles tend to be the same length.
- How could you select a random sample of aisles?
- Within an aisle, how would you choose the left side or the right side?
- Within an aisle side, how could you randomly select a shelf?
- Within a shelf, how could you randomly select a 2 foot section?

(f) Suppose you then take every unique product in each of the 2 foot sections selected for several aisles and combined these together as your sample. Would finding the items for the sample selected in (e) be any easier than the previous two plans? Explain.

Def: The sampling method you described in (e) is a type of *multistage cluster sample*. At the last step we took a *cluster* of all items on the shelf. Note, this cluster should have different products. In a *multistage sampling plan,* random samples or even systematic samples are taken at each stage and then the units selected are combined to form the sample.

One advantage of multistage cluster sampling is that we don't need a list of all items in the store and we know exactly where the item is in the store.

(g) In the above plan, does every product have an equal chance of being selected? Explain.

(h) Is it possible that the above sampling methods would fail to select a non-food item (e.g., laundry detergent, toothpaste) in one sample? Do you think prices of non-food items might differ from prices for food items? Explain.

Def: Another sampling method is a *stratified random sample* where the population is first divided into similar groups of individuals, called *strata*. Then a separate random sample is chosen within each stratum and these units are combined to form the sample. One way to stratify is *proportionally*, where the sample percentages of items in each strata are chosen to match the percentage of the population that is in each strata.

Discussion: One distinction between cluster sampling and stratified sampling is that the "groups" you work with in a cluster sample are not expected to be homogeneous, but with a stratified sample, the separate "groups" are chosen to be more homogeneous. Note, this is similar to the idea of *blocking* (Chapter 1) in that we first group the items that are similar and we expect there to be differences between the groups. In this case, we are taking an extra step to ensure that our sample is representative of the population, by forcing our sample to match this characteristic (e.g., the breakdown of food and non-food items) of the population.

(i) Describe how you might conduct a stratified random sample of 30 items, stratifying the food and non-food items in store L. (Assume about 75% of the products are food items.)

All of the above methods fall under the class of *probability samples*. The key point to remember is that some random mechanism is responsible for selecting the units in the sample, not the researcher and not the observational individuals themselves. Probability samples have a large advantage in that they help ensure that the samples selected are representative of the larger population. We may sometimes take additional steps to make the sampling method more efficient or convenient or when we have more information about the population (e.g., percentage in different strata) that we want to take into account.

In analyzing the results for these more complicated sampling plans, we need more complicated techniques. However, often we will initially assume the sample is behaving like a simple random sample as we explore the results.

In the next few sections you will see that the second key advantage of a probability sample is that we can apply probability methods to determine how much we can expect the sample results to deviate from the population parameters, by chance.

Practice:

3-3) Sampling Words (cont.)
Suppose we want a sample of 10 words from the Gettysburg Address and want to look at the length of words. Describe how you might carry out each of the following sampling plans:
(a) Systematic sample

(b) Multistage cluster sample

(c) Stratified random sample (be sure to indicate what variable you would stratify on)

3-4) Sampling Airline Passengers
Suppose an airline company wants to interview their passengers this year to assess their satisfaction with the service. Explain how you might carry out a multistage cluster sample so that you did not need a list of all passengers on all flights.

3-5) Sampling Apartment Residents
Suppose we have a series of 10-story apartment buildings and we wanted to interview residents about their willingness to buy extra insurance (e.g., for floods, theft). Give an example where *periodicity* could cause problems for a systematic sampling plan.

Investigation 3-3: Sampling Variability of Sample Means

The two previous activities illustrated proper ways to use probability to obtain samples that can be assumed to be representative of the population. In this investigation, you will further explore the properties of such simple random samples.

In Investigation 3-1, each student generated a random sample of 5 words from the Gettysburg Address and you looked at the distribution of the sample mean word length for these samples. To thoroughly investigate the long-run properties of this sampling method, we would need to examine all possible samples of size 5 from this population.

Def: The *sampling distribution of a statistic* is the distribution of the sample statistic for all possible samples (of the same size) from the population. It is very important that you distinguish between the *sample*, the *population*, and the *sampling distribution* in a given context.

(a) For the Gettysburg Address investigation where you examined the average length of words, describe in words the population, a sample, and the sampling distribution.

Population: Sample:

Sampling Distribution (shape, center, spread):

(b) How many different samples of size 5 are possible from this population of words? [*Hint*: Recall counting methods from Chapter 1.]

Instead of listing all of these possible samples (more than 11 billion of them!) and calculating the sample mean for each sample to determine the exact sampling distribution, we will use an applet to generate many random samples for us and display an *empirical sampling distribution* of the sample means. This empirical sampling distribution will approximate the exact sampling distribution.

(c) Open the "Sampling Words" applet.
You will notice that the applet displays the population of word lengths in the top right, along with the long/short words and the nouns/not nouns in the population. For now, uncheck the boxes next to "Show Long" and "Show Noun".

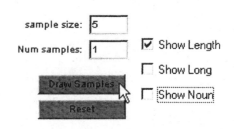

Describe the distribution of word lengths in this population (shape, center, and spread).

(d) Specify 5 as the sample size and click the Draw Samples button. Report the five words in your sample.

(e) What is the average word length ($\bar{x}$) in this sample? Note that the applet reports this value in the top dotplot (the black arrow) and also places this value in the bottom dotplot.

First $\bar{x}$:

(f) Click Draw Samples again; did you get the same sample of words? Did you obtain the same sample mean word length? What is the probability that any particular sample of 5 words will occur?

Second $\bar{x}$:

Probability:

(g) What is the average of these two sample means?

Note that the applet displays the average of the two sample means by a red arrow in the bottom dotplot and reports the value in the upper right corner of the window for that dotplot.

(h) Now change the number of samples ("Num samples") from 1 to 98. Click Draw Samples. The bottom dotplot displays the distribution of these 100 sample means. Now uncheck the animate box and change the number of samples to 300. What are the observational units and variable for the bottom dotplot? Describe the resulting empirical sampling distribution, remembering to comment on center, spread, shape, and outliers. Also make sure you report the mean and the standard deviation.

Observational units:

Variable:

Shape:

Center, mean $\bar{x}$ value:

Spread, standard deviation of $\bar{x}$ values:

Visual outliers?

(i) How does your distribution compare to your neighbor's? Does there appear to be a consistent pattern to how the average length of words varies from sample to sample?

(j) Change the sample size from 5 to 10 and the number of samples from 300 to 400 and click Draw Samples. The previous distribution of 400 samples means will turn green, and the new empirical sampling distribution of 400 sample means will be represented in black. How do the two distributions compare? [Comment on center, spread, and shape.] In particular, how do the mean and standard deviation of the distributions compare?

(k) Do these sample means (with $n=10$) tend to lie closer or further from the population mean than for $n=5$?

(l) Recall the average length of words you found when you took your nonrandom sample of 10 words in Investigation 3-1. Roughly how often (in what proportion of samples) did you obtain a sample mean at least that extreme when the samples are randomly selected?
[*Note*: If you click on a sample mean in the sampling distribution graph, the applet will show you the corresponding sample in the top window and the value of the mean for that sample.]

(m) Suppose that Scott tells you that he obtained a sample mean of 6.7 letters. Would this be a surprising result if he selected his sample of 10 words randomly? Explain.

(n) Repeat (m) for Kathy, whose sample mean was 4.8 letters.

(o) Repeat (n) for Scott, but suppose his sample had only $n=5$ words.

(p) Which scenario has the largest (in absolute value) z-score among Scott ($n=10$), Kathy, and Scott ($n=5$)? [*Hint*: Use the empirical standard deviations given by the applet.]

Discussion: Once again you see the fundamental principle of *sampling variability*. Fortunately, this variability follows a predictable pattern in the long run. You should see that the average of the sample means is approximately equal to the population mean, $\mu = 4.29$, as we would expect since the simple random sampling method is unbiased. [The mean of your empirical sampling distribution may not be exactly equal to μ since you only took 400 samples, instead of all possible samples of size 5, but we are not consistently overestimating the population mean as we did with the nonrandom samples.] This consistent pattern helps us to decide when we might have a surprising value for the sample statistic. Of course our level of surprise will depend on the sample size since you also saw that the amount of sampling variability in the sample mean decreases when we increase the sample size. Sample means that are based on larger samples will tend to fall even closer to the population mean μ and there is less variability among the sample means. So first, select *randomly* to avoid bias, and then if we can *increase* the sample size, this will improve the *precision* of the sample results.

Study Conclusions:
The calculations in (l)-(o) are empirical *p-values*. In Chapters 1 and 2, you used the p-value to decide if there was a statistically significant difference between two groups, and you learned that "statistically significant" means "unlikely to happen by chance alone." Here, you are considering just one variable and one group, but are interested in whether the sample result is significantly different from the population value, where the sample is a subset of the observational units in the population. If you obtained a sample statistic that was so different from the population parameter that you don't believe it arose by chance alone, then you have evidence that your sample was not randomly selected from the specified population. A small p-value does not guarantee that the sample result did not happen just by chance, but it does allow

you to measure how unlikely such a result is. The size of the p-value will directly depend on the amount of sampling variability. The same sample mean will seem more unusual if it came from a larger sample than a smaller sample (larger z-score, smaller p-value) as you saw in (m) and (p).

You should keep in mind that the above investigation replicated the simple random sampling process. However, the results for other sampling methods will often yield similar findings in that they will produce unbiased statistics that become more precise with larger samples. While we should employ more complicated analysis techniques to correspond to these more complicated sampling methods, in this text we will often assume that those samples are behaving like simple random samples. You can learn more appropriate techniques in a subsequent course on sampling design and methods.

Practice:

3-6) Stock Portfolios
Use the above observations to justify why a stock broker may prefer to diversify a portfolio by investing in several stocks rather than in only 1 or 2 stocks.

3-7) Roulette
In the game of Roulette, your expected winnings per spin is about minus 5 cents. Explain why a casino goes to extreme lengths (e.g., no windows, free drinks) to encourage you to play the game for long periods of time.

3-8) Sampling Words (cont.)
Suppose your population consisted of just the words: I, pledge, allegiance, to, the
(a) Produce a dotplot of the word lengths in this population.
(b) What is the average length of words in this population? What symbol would you use to represent this average?
(c) If we took samples of size $n=2$ from this population, how many different samples are possible?
(d) List the samples in (c) and calculate the sample mean for each sample.
(e) Produce a dotplot of the distribution of all possible sample means for samples of size $n=2$ from this population, the *exact sampling distribution*.
(f) Calculate the mean of the sampling distribution of sample means in (d). How does it compare to the population mean determined in (b)?
(g) How does the variability of the sampling distribution of sample means in (d) compare to the variability of the population in (a)?

Investigation 3-4: Sampling Variability of Sample Proportions

Return again to the Gettysburg Address example and to the variable "whether or not the word is long (more than 4 letters)." This was a categorical variable and we looked at a bar graph to display the results of this variable for your nonrandom sample. The sample statistic can be calculated as the proportion of long words in the sample, denoted by $\hat{p}$, and the corresponding population parameter would be the proportion of long words in the population, denoted by π. Recall that there were 99 long words and 169 short words in the population (so $\pi = 99/268 = .369$). In this investigation you will focus on the *sampling distribution of sample proportions*.

(a) If we repeatedly take simple random samples of size $n=5$ from the population of words in the Gettysburg address and calculate a $\hat{p}$ for each sample, where do you expect the distribution of these sample proportions to be centered? Explain. Is this distribution the sample, the population, or the sampling distribution?

(b) How do you expect this sampling distribution to change if we instead repeatedly took random samples of size $n=10$ from the population of words in the Gettysburg Address?

To verify your conjectures, we will again turn to technology to simulate the long-run pattern of the sampling distribution of sample proportions. Return to the "Sampling Words" applet and this time, check only the "Show Long" box.

(c) Use the applet to select 400 samples of size $n=5$ from the population. Describe the shape, center, and spread of this empirical sampling distribution.

Shape:

Center, average $\hat{p}$ value:

Spread, standard deviation of $\hat{p}$ values:

(d) Change the sample size from *n=5* to *n=10* and click Draw Samples. Describe the shape, center, and spread of this empirical sampling distribution and how this distribution compares to that in (c). Are these observations consistent with your predictions in (a) and (b)?

Shape:

Center, average $\hat{p}$ value:

Spread, standard deviation of $\hat{p}$ values:

Differences from distribution in (c):

Comparison to predictions:

***Discussion*:** You should again see that when we take random samples, the distribution of the sample proportion centers at the value of the population proportion π (the sample proportion is an *unbiased estimator* of the population proportion). Furthermore, this distribution is less variable when we use larger sample sizes. One difference between the distribution of the sample means and the distribution of sample proportions is the *granularity*, or discreteness, of the sample proportions which is less severe with larger samples. For example, in the *n=5* case, there are only six possible values for $\hat{p}$: 0, .2, .4, .6, .8, and 1.

Another key difference between working with means and proportions is that it is more straight-forward to determine the exact sampling distribution by counting how many samples lead to these few different sample proportion values.

(e) How many different possible samples of size *n=5* are possible? What is the probability of each sample occurring? How many of these samples contain 5 long words?

Number of possible samples:

Probability of each sample:

Number of samples containing 5 long words:

(f) Use your answers in (e) to determine the probability of obtaining 5 long words in a randomly selected sample of size *n=5*.

Discussion: Since we are using simple random samples, each sample has the same probability of occurring. Therefore, to determine the probability of a particular outcome, e.g., 5 long words, we count the number of samples with 5 long words and divide by the total number of possible

samples: $\dfrac{\dbinom{99}{5}\dbinom{169}{0}}{\dbinom{268}{5}} = .0064$, so it would be very unlikely to obtain all long words in a sample

of 5 words. Note that these probabilities will be the same whether we consider the *number* of long words or the *proportion* of long words. However, if we work with the number of long words, we obtain a very familiar probability model.

(g) Let the random variable X represent the number of long words in a randomly selected sample of 5 words from this population. Can X be considered a hypergeometric random variable? Explain.

(h) As you did in Activity 1-12, enter the values 0-5 in C1 and then have Minitab determine the Hypergeometric probabilities associated with those values, storing the results in C2 and using 268 as the population size, 99 as the number of successes in the population, and 5 as the sample size. You can convert the counts to proportions as well:

```
MTB> let c3=c1/5
MTB> name c3 'phat'
```

To examine a display of this probability distribution, type:

```
MTB> plot c2*c3;
SUBC> project.
```

Note: if you double click on the *y*-axis, you can set the scale range minimum to be 0.

Would you consider this distribution "skewed"? If so, to the right or to the left? How does this distribution compare to the empirical sampling distribution from (c)?

(i) Calculate the expected value. $E(X) = \Sigma x P(X=x)$, of this distribution either by hand or type:

```
MTB> let c4=c3*c2
MTB> sum c4
```

Report this expected value. How does this expected value compare to the average $\hat{p}$ value you found in (c)?

Discussion: The exact sampling distribution of the sample proportion of long words for *n*=5 is:

$\hat{p}$	0	.2	.4	.6	.8	1
Prob	.098	.293	.345	.201	.057	.006

which has a very slight skew to the right. This probability distribution tells you the possible values of $\hat{p}$ and the exact probability for how often each $\hat{p}$ value will occur. Your empirical sampling distribution in (c) should provide a reasonable approximation to this exact sampling distribution, and the approximation would improve if you took a larger number of samples. The expected value of $\hat{p}$ turns out to be .369, which is not coincidentally the value of the population proportion π.

(j) Use Minitab to determine and display the exact sampling distribution of $\hat{p}$ when *n*=10.

$\hat{p}$	0	.1	.2	.3	.4	.5	.6	.7	.8	.9	1.0
Prob											

(k) What is the expected value of the above probability distribution?

(l) How does this exact sampling distribution compare to the empirical sampling distribution shown in the applet?

(m) How does this compare to the sampling distribution when *n*=5?

(n) What is the probability of randomly selecting 10 words and finding 10 long words? How does this probability compare to that calculated in (f)? Explain why this comparison makes sense.

   P($\hat{p}$ = 1):

   Comparison to (f), when *n*=5:

STATISTICAL INFERENCE

(o) Suppose we did not know how many nouns were in the Gettysburg Address but we suspected that there were 50 nouns ($\pi = .187$). Suppose that a student takes a random sample of 10 words and obtains one noun. Use the hypergeometric distribution to decide if obtaining at most 1 noun ($\hat{p} \leq .10$) would be a surprising outcome if the sample were drawn randomly from a population with 50 nouns.

(p) Repeat (o) to determine the probability of a random sample of 10 words containing 5 or more nouns. [*Hint*: Recall the complement rule: $P(X \geq x) = 1 - P(X \leq x-1)$.]

Summary: With a categorical variable, if we sample from a finite population without replacement, we are able to compute the probability for a particular value of $\hat{p} = x/n$ exactly using the hypergeometric distribution since we just need to count how many samples have x successes and $n-x$ failures. When we know the value of the population proportion, this calculation tells us the probability of obtaining particular values of $\hat{p}$ from that population. When we don't know the value of the population proportion but we know the sample is selected at random, we can use the hypergeometric distribution to compute a p-value which allows us to make decisions about a conjectured value of the population proportion as in (o) and (p).

While the distribution of sample proportions is very analogous to the distribution of sample means (centered at the population parameter, less variable with larger sample sizes), the exact probability distribution for the sample mean is extremely cumbersome to calculate. This is analogous to what you did in Chapter 1, using the hypergeometric probability distribution for the exact randomization distribution with categorical variables (Fisher's Exact Test) but relying on a simulated randomization distribution for a quantitative response variable (Chapter 2). In the next chapter you will learn some additional probability models that will help us model the behavior of the different sampling distributions when the exact distribution cannot be determined analytically.

Practice:

3-9) Sampling Words (cont.)
(a) Return to the "Sampling Words" applet and generate an empirical sampling distribution (400 samples) for the mean length of words and the proportion of long words with $n=10$. Describe the shape, center, and spread for each distribution, reporting the values of the average and standard deviation of the empirical sampling distribution.
(b) Click Reset. Use the pull-down menu at the bottom left of the screen to change from "address" to "fouraddreses". Your population now consists of four copies of the Gettysburg Address, so the population size is four times larger than before, with the same population parameter values. How do you think the empirical sampling distributions will change if we again look at samples of $n=10$ words? Explain.
(c) Click Draw Samples. Comment on how the shape, center, and spread for these empirical sampling distributions compare to those in (a).
(d) Did the change in population size lead to large differences in the behavior of the sampling distributions? Is this consistent with your conjecture in (b)?

3-10) Sampling Words (cont.)
Recall that in part (o) in Investigation 3-4, you found that the probability of obtaining at most one noun in a random sample of 10 words from the Gettysburg Address, assuming 50 nouns in the population, is .414.
(a) Continue to suppose that there are 50 nouns in the population ($\pi = .187$) but a student obtains a random sample of 100 words which contained 10 nouns ($\hat{p} = .10$). Use the hypergeometric distribution and Minitab to decide if obtaining at most 10 nouns would be a surprising outcome. How does this p-value compare to that of part (o) in Investigation 3-4? Explain why this relationship makes sense.
(b) Suppose we were sampling from 4 copies of the Gettysburg Address so that the population size was 1072 words. Suppose we thought there were 200 nouns in the Gettysburg Address ($\pi = .187$) but our random sample of 10 words only contained 1 noun ($\hat{p} = .10$). Use the hypergeometric distribution to decide if obtaining at most 1 noun would be a surprising outcome. Comment on how this probability compares to that from part (o) in Investigation 3-4.
(c) Now redo part (a) for this larger population. How does this p-value compare to that from part (a)?
(d) Does the population size appear to affect these p-value calculations considerably? Explain.

STATISTICAL INFERENCE

In the previous activities you saw that you can study the sampling distribution of a statistic and make probability statements about likely (or unlikely) values for the statistic to take. But we *knew* the distribution of the entire population in those activities.

Now you will see how this knowledge of a sampling distribution can help you to make inferences about an unobserved population based on a sample. The reasoning process is a familiar one: we'll ask whether the sample results are very surprising given a hypothesis about the population.

Investigation 3-5: Freshmen Voting Patterns

A student project group (Norquest, Bayer, and McConville, 2004) wanted to determine whether there was a preference for the primary presidential candidates (George Bush and John Kerry) among first-year students at their university. They believed that these younger students would be more liberal and more inclined to vote for the democratic candidate. They took a sample of 30 students from the 705 first-years at their school by assigning the first-year residence halls a number between 1 and 5 (inclusive) and rolling a die to select a residence hall. Then from the selected residence hall they took every seventh room to be part of the sample. The survey was distributed by going from room to room with the surveys, giving each resident of the room a copy of the survey to fill out, and immediately collecting the surveys after they were completed. If one of the residents was not in the room at the time of the first visit, the group repeatedly returned to the room until the person was contacted. The surveys were anonymous and the group members did not look at the completed surveys until they were randomly scrambled. The survey contained three questions, the first question was whether the student planned to vote for Kerry or Bush, and the other two questions concerned other social issues. For half of the surveys Kerry was listed first and for half of the surveys Bush was listed first.

(a) Identify the observational units and response variable in this study. Is the response variable quantitative or categorical?

    Observational units:

    Variable: Type:

(b) Identify the sample, the population, and the sampling frame used in this study.

    Sample:

    Population:

    Sampling frame:

(c) Was the sampling method used a simple random sample, a stratified sample, a systematic sample, a multistage cluster sample, or some combination? Do you think the sample selected will be representative of the population of interest? Explain.

> **Def:** *Nonsampling errors* can occur even after we have a randomly selected sample. They are not associated with the sampling process but rather with sources of bias that can arise after the sample has been selected. Sources of nonsampling errors in surveys include the word choice in survey questions, dishonest or inaccurate responses by respondents due to sensitive questions or faulty memory, the order in which questions appear, a leading tone and appearance of interviewer.

(d) Identify some precautions taken by these students to avoid nonsampling errors.

Unlike the Gettysburg Address investigation, we do not know the distribution of candidate preferences in the population of 705 first year students. But we can use the sample results to "test" the plausibility of various population parameter values by asking how surprising the sample results would be for these values.

> If we conjecture a value for the population parameter, but then find that our sample results are unlikely to occur by chance alone (the p-value is small) when that conjecture is true, we will have evidence against our conjecture about the population. In this case, we will conclude that our sample data provide *statistically significant* evidence against the conjecture. As before, the smaller the p-value, the stronger the evidence against the conjecture.

In this study the sample results were that 22 of the 30 first-years planned to vote for John Kerry, while 8 planned to vote for George Bush.

(e) Construct a bar graph to display the results for this sample and write a one-sentence summary of what the graph reveals.

(f) Suppose 352 first-year students at this school (essentially 50% of the population) plan to vote for Kerry. Would the sample result obtained by these students be surprising? Explain.
[*Hint*: Find the probability of observing 22 or more who favor Kerry in a random sample of 30 students from such a population.]

(g) The student project group suspected that about 2/3 of the first-year students at this school are planning to vote for Kerry. If 470 first-year students at this school plan to vote for Kerry (2/3 of this population), would the sample result obtained by these students by surprising? Explain.

(h) Which seems more plausible for the number of Kerry supporters in this population, 470 or 352? Explain.

Study Conclusions: In carrying out a sample survey, we need to be considered with both *sampling errors* and *nonsampling errors*. These students avoided sampling errors by carrying out a multi-stage cluster sample, using a systematic sample within the cluster (leaving only *random sampling error*). This is much preferred to a convenience sample, where they would have simply asked their friends "at random" who they planned to vote for. However, this sampling plan could still be problematic if there are differences in the voting preferences among the dorms. They maintained the randomness in the selection process by repeatedly visiting the rooms until the person selected was contacted. This is important as students who are not in their rooms may have different political views than students who are easily found in their rooms. These students also took several precautions to avoid nonsampling errors such as rotating the names of the candidates, including other distractor questions on the survey, and also ensuring confidentiality on the surveys to avoid pressuring the respondents to vote in a particular way.

Even though this was not technically a simple random sample, we can assume it behaves like a simple random sample and then use the hypergeometric distribution to "test" some conjectures about the population. If roughly half the population planned to vote for Kerry (352 supporters, π = .5) it would be very surprising for a random sample of 30 respondents to have at least 22 indicate that they planned to vote for Kerry (p-value=1-.993=.007, see graph below on the left). On the other hand, if 2/3 of the population planned to vote for Kerry (470 supporters, π = 2/3) then a sample result of 22 or more successes out of 30 is not at all surprising (p-value = 1-.719 = .281, see graph below on the right). With such a large p-value, we do not have significant evidence against the belief that there are 470 Kerry supporters in this population. Based on this analysis, 2/3 is a much more plausible value than 1/2 for π, the proportion of Kerry supporters among first year students at this school, as the observed sample result is a much more typical value for that distribution.

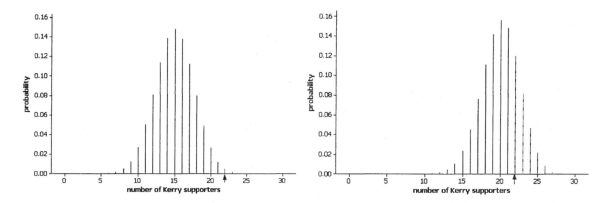

Discussion: We have been very careful to talk about "plausible" values of π instead of "probable" values of π. Most statisticians prefer not to use language such as "π is probably 2/3" because π has a fixed value (just unknown to us exactly) and is not changing. So our "long-term relative frequency" definition of probability does not apply to statements about π. We can make probability statements about the sample statistics, because these are random, but not about the population parameter.

Practice:

3-11) Freshmen Voting Patterns (cont.)
Reconsider Investigation 3-5. Let k equal the number of Kerry supporters in the population of all 705 first-year students at this school.
(a) Determine the largest value of k such that the probability of getting 22 or more Kerry supporters in a random sample of 30 freshmen would be less than .05. (Provide several calculations and an explanation to support your answer.)
(b) Determine the smallest value of k such that the probability of getting 22 or fewer Kerry supporters in a random sample of 30 freshmen would be less than .05.
(c) Explain the sense in which the values of k between these two extremes are plausible values for the number of Kerry supporters in the population in light of the sample results.

3-12) Cycling Accidents
A recent project of the Boston Metropolitan Area Planning Council was a study of bicycle-motor vehicle accidents in the Boston area (along Route 128, a major beltway encircling 35 communities), http://www.bikexprt.com/research/ctps/cover.htm. In particular, the researchers wanted to decide whether more accidents occurred on Fridays than was expected if the accident rate was the same for all seven days of the week. As almost 2000 accidents had been reported for 1979 and 1980, the researchers decided to study a sample of the reported accidents. The selection of accidents was done using a computer printout provided by the Massachusetts Department of Public Works of all bicycle-motor vehicle accidents occurring in the study area during 1979 and 1980. One in four accidents was selected for review. Of the 512 accidents, 87 occurred on Fridays. [There appears to be an inconsistency in the report - you can assume there were 2050 accidents reported.]
(a) Identify the observational units, the population, the sample, and the sampling frame for this study.
(b) Use the hypergeometric probability distribution to determine the probability of observing at least 87 accidents on Fridays if $1/7^{th}$ of all 2050 accidents occurred on Fridays (293).
(c) Is the p-value you calculated in (b) small enough to convince you that 293 is not a plausible value for the number of Friday accidents in the population? Explain.

Investigation 3-6: Comparison of Sampling Methods (Minitab Exploration)

In this exploration, you will use a Minitab macro to examine some of the properties of the sampling methods described in Investigation 3-2.

The file `shoppingPop.mtw` contains prices for 144 foods. Treat this as your population.
(a) Calculate the population mean and describe the shape and variability of the population.

(b) Write a Minitab macro to take 1000 samples, each containing 14 products, from this population, calculate the mean for each sample, and accumulate the results in C4. Describe the shape, center, and spread of this sampling distribution.

(c) Column 2 contains information on the two strata (groups) – food items and non-food items (where we have included spices and household goods). Use Minitab's "unstack" command to separate these two groups:

```
MTB> unstack c2 c5 c6;     #places the results into columns 5 and 6
SUBC> subs c3.             #indicates that column 3 has the group identifiers
```
Note: These steps can be reversed using Minitab's "stack" command
```
MTB> stack c5 c6 c7;       #places the results into column 7
SUBC> subs c8.             #keeps track of which group they came from.
```

After running the unstack command, how many items are there in each group (e.g., `MTB> info c5 c6`)? What percentage of the population has been classified as a food item?

(d) Does there appear to be a difference in the price distributions between these two groups? Support your answer with appropriate numerical and graphical summaries. [*Hint*: Boxplots are very helpful here.]

(e) Now write a Minitab macro to take 1000 stratified samples of 14 products, with the stipulation that each sample must have 4 non-food products (from C5) and 10 food products (from C6). [Note that this food/non-food breakdown roughly matches the percentage breakdown in the population.] The macro should stack the resulting samples into C7, and then calculate the mean for each combined sample and store the results in C9. Describe this sampling distribution and compare it to the one generated in (b).

(f) Which sampling method, the simple random samples or the stratified samples, was/were unbiased? Which had less variability? Explain why these answers make sense.

(g) Create your own grouping variable (e.g., the first 72 items and the second 72 items) and explore a stratified sampling method using these strata. Does this stratification appear to improve the precision of the sample means? Explain.

SECTION 3-2: SAMPLING FROM A PROCESS

In this section, we will again explore the idea of sampling variability. Instead of sampling from a population, we will consider sampling from a *process*. We will focus on binary variables.

Investigation 3-7: Do Pets Look Like Their Owners?

Researchers recently conducted a study to see if people could match up the picture of a pet with a picture of its owner (Roy and Christenfeld, 2004). For example, open the "Pet Owner" webpage. You will be presented with the picture of a cat owner and three cats.

(a) Which cat do you think belongs with this cat owner?

(b) If there really was no connection between the appearance of a cat and its owner, what is the probability that you will make the correct match?

(c) How does the probability given in (b) change from person to person if each person in your class is actually just guessing which cat belongs to this owner?

(d) Suppose the person next to you guesses correctly, does this make you more or less likely to guess correctly?

Discussion: In the previous section we defined samples from a finite population. In this investigation, we don't have a well-defined population of observational units. Instead, we will consider this a sample from a *process*. The parameter of interest is the probability of success and we will continue to use the symbol π to represent this unknown parameter.

In this setting, let π represent the probability that a person guesses correctly. If everyone is simply guessing which cat owns the person shown, then π is equal to 1/3 for each guesser. Furthermore, we consider these observations *independent* in that whether or not one person guesses correctly does not affect the probability that another person guesses correctly. A counter example to this is the random babies scenario, where knowing whether or not one mother has received the correct baby directly affects the probability that the next mother will receive her correct baby.

> **Def:** A *Bernoulli process* has the following characteristics:
> - Each observational unit or *trial* falls into one of two categories (call one category a "success" and one a "failure")
> - The probability of success (denoted as π) is the same for each observational unit
> - The results for the observational units are independent.

The prototypical example of a Bernoulli process is the tossing of a coin and determining whether it lands heads or tails. We don't have a finite population of coin tosses, but an infinite *process* from which the coin toss results are generated. Similarly, the outcome of a die roll can be considered a Bernoulli process. Suppose we consider rolling a 2 as a success. Then the probability of success is 1/6 for each roll of the die, and the results of one die roll do not affect the results of another die roll. Notice we are not saying P(success)=P(failure), only that P(success) is the same for each observational unit (e.g., each die roll, each coin toss).

The idea of independence of observational units is slightly different than the use of the term *independence* in Chapter 1. There, we looked at independence of *variables* – knowing the outcome of one variable did not give us additional knowledge about the outcome of a second variable. Here, we are considering the *independence* of different *trials* – knowing the outcome of one trial does not change the probability we assign to obtaining a success on the next trial.

Matching the owner to a cat can be considered a Bernoulli process since, assuming everyone is guessing, $\pi = 1/3$ for each person, and the trials (different attempts to match the cat to its owner) are independent.

(e) Let Y represent the random variable that assigns a 1 to a correct guess and a 0 to an incorrect guess. What is the probability distribution of Y? [List the possible outcomes and the probability of each outcome.]

(f) What was the empirical distribution of Y obtained by your class?

(g) What was the average value of Y observed in your class?

(h) Recall that for a discrete probability distribution, we can determine the expected value of the random variable through the formula $E(Y) = \Sigma y_i P(Y=y_i)$. Calculate this quantity and compare your class average to this theoretical expected value.

Discussion: When we have a Bernoulli process we can define a random variable for each trial by denoting a success by "1" and failures by "0." The probability distribution is simply $P(Y=1)=\pi$ and $P(Y=0)=1-\pi$, and the expected value of Y is then $E(Y)=1\times\pi + 0\times(1-\pi) = \pi$. The only parameter of this probability distribution is π.

We will now move from considering a 0/1 random variable for each trial to studying a random variable that counts the number of "successes" in a fixed number of trials.

Investigation 3-8: Pop Quiz! (Probability Exploration)

(a) You must now take a surprise 5 question quiz. Each question has 4 multiple choice options, A, B, C, or D. Write down your answers to the 5 questions.

<u>A</u>   <u>A</u>   <u>D</u>   <u>C</u>   <u>B</u>

(b) Based on your instructor's answer key, how many of your responses are correct?

(c) Does this situation have the characteristics of a Bernoulli process?

- How are you defining a "success" and a "failure"?

- If π denotes the probability of a correct answer to a single question, what is the value of π for question 1? Question 2? Question 3? Question 4? Question 5?

- Does your result for one question affect the result of another question?

(d) Let the random variable X represent the number of correct answers a guesser gives in these 5 questions. What are the possible values of X? Will everyone in your class observe the same value of X? Explain.

　　　Possible values:

　　　Will X vary?

(e) Record a tally of the results for your class:

Number correct	0	1	2	3	4	5
Number of students						

(f) Based on your class results, what is your empirical estimate for the probability that someone gets exactly one question correct, $P(X=1)$?

Practice:

3-13) Binomial or Not?
Explain why the following are *not* Binomial random variables.
(a) Let X = number of computer solitaire games you play before your first win.
(b) Let Y = time you wait in line at the convenience store each time you visit.
(c) Let Z = number of long words in a sample of 10 words from the Gettysburg address.

3-14) ESP
A standard ESP test asks subjects to identify which of four shapes (e.g., circle, square, diamond, waves) is on the back of a card. Suppose that a subject takes a test with 25 of these cards. Let the random variable C denote the number of correct identifications.
(a) Does C follow a binomial distribution? If so, justify each criterion. Also specify the parameters of the distribution, assuming that the subject is guessing.
(b) What is the probability that the subject would get 10 or more correct?
(c) How many identifications would the subject have to get correct in order for the probability to be less than .01 that a guesser would get at least that many correct? [*Hint*: State the desired inequality and use the "inverse cumulative probability" option.]

3-15) Binomial Distributions
Use the applet or Minitab to examine the following binomial distributions
(a) Binomial (n=25, π = .10)
(b) Binomial (n=25, π = .50)
(c) Binomial (n=25, π = .90)
Describe how the shape, center, and spread of these three distributions compare.

SECTION 3-3: EXACT BINOMIAL INFERENCE

In the previous section, you calculated probabilities from a specific random process where the value of π, the probability of success in the process, was known. In this section you will instead focus on making inferences from your sample result to the (unknown) value of π. You will apply the binomial distribution to determine p-values in order assess the strength of evidence against particular conjectured values of the probability of success in a process. You will also learn a more formal structure for carrying out a *test of significance*. It's important that you keep in mind the general interpretation of a p-value even as we introduce this additional structure and terminology. You will also learn how to specify an interval of plausible values for the probability of success.

Investigation 3-9: Water Oxygen Levels

Scientists often monitor the "health" of water systems to determine the impact of different changes in the environment. For example, Riggs (2002) reports on a recent case study that monitored the dissolved oxygen downstream from a commercial hog operation. There had been problems at this site for several years (manure lagoon overflow), including several fish deaths in the previous 3 years just downstream of a large swale through which runoff from the hog facility had escaped. The state pollution control agency decided to closely monitor dissolved oxygen downstream of the swale for the next 3 years to determine if the problem had been resolved. In particular, they wanted to see how often the dissolved oxygen level in the river was less than the 5.0 mg/l standard. A measurement with a lower level was considered "noncompliant." Sampling was scheduled to commence in January of 2000 and run through December of 2002. The monitors took measurements at a single point in the river, approximately six tenths of a mile from the swale, once every 11 days.

(a) Identify the observational units in this study.

(b) Since the dissolved oxygen measurements were taken in the same location at fixed time intervals, would you consider this a simple random sample, a systematic sample, a multistage cluster sample or a stratified random sample? Explain.

(c) In this study, we will consider measurements of the dissolved oxygen for the river to be from a process where π represents the probability of a noncompliant measurement ("success"). Do you believe this sampling method will lead to samples with characteristics similar to the overall process? Explain. What precautions should be taken when interpreting results of this study?

Discussion: These last two investigations have given you practice in carrying out a test of significance using the binomial distribution to find the p-value. Note that the inequality used in calculating the p-value is determined by the alternative hypothesis. In both of these examples, the alternative hypothesis conjectured that the process probability π was greater than the benchmark value, so the p-value was calculated as the probability of obtaining *at least* as many successes as in the sample. If the alternative hypothesis had been that π was less than the comparison value, then you would have calculated the p-value as the probability of obtaining *at most* this many successes as in the sample.

It is important that you always consider whether the underlying binomial model is appropriate in a particular situation. You may sometimes calculate "hypothetical p-values" where there is no real source of randomness in the study, but you still want a measure of how surprising the observed result is. Just be very cautious in not generalizing these results to a larger process. In the next few activities you will consider more sophisticated binomial probability calculations.

Practice:

3-16) Stating Hypotheses
For each situation below, define the parameter of interest and state the null and alternative hypotheses….
(a) …for testing whether a subject has ESP (recall Practice 3-14).
(b) …for testing the students' conjecture that a majority of freshmen at their school were planning to vote for Kerry (recall Investigation 3-5).
(c) … for testing whether employees are taking sick days on Mondays and Fridays at a higher rate than would be expected by chance (as once conjectured by Dilbert's pointy-haired boss).

3-17) Monitoring Mortality (cont.)
In the previous study, the researchers defined "excess mortality" for transplantations to be 20%. Repeat the previous analysis based on the sample of 361 surgeries, using .20 as the hypothesized value of π.
(a) State the new null and alternative hypotheses (in symbols and in words).
(b) Assuming this new null hypothesis is true, how would the above probability distribution change? What is the expected value of X, the number of deaths in 361 surgeries?
(c) What is the p-value for this new null hypothesis based on the 71 successes?
(d) Is this p-value larger or smaller than the one above? Explain why this relationship makes sense.
(e) Do you reject or fail to reject the null hypothesis at the 5% level?
(f) Write a one-sentence summary to these researchers of whether there is significant evidence of "excess mortality" at this hospital.

3-18) Flat Tires
A campus legend tells the story of two students who were late returning from the weekend for their Monday Chemistry midterm. They explained to the professor that they had a flat tire on their drive back into town and hoped he would give them a make-up exam. The professor agreed (a bit to the students' surprise) and they returned the next day to take the make-up exam. He sent them to different rooms and they found the first question, worth 5 points, pretty straight-forward.

They turned the exam sheet over and found the second question, worth 95 points, was "which tire was it?"

(a) Suppose you found yourself in this situation but had made up the story about the flat tire. Which tire would you indicate on the exam?

(b) Suppose we keep track of how often people specify the "right front" tire. Let π denote the probability of a randomly selected person specifying the right front tire. If people are guessing tires at random, what is the value of π?

(c) State the null and alternative hypotheses for deciding if people are more likely to pick the right front tire than the other tires.

(d) In a class of 52 students, we counted 22 people who specified the right front tire. Do we have a Bernoulli process here?

(e) Use the Binomial Simulation applet to simulate the responses for 52 people in a large number of different samples where $\pi = .25$. How often did the simulation produce at least 22 people specifying the right front tire? Does this appear to be an unusual outcome when people are guessing at random? Explain. Would you consider this empirical p-value convincing evidence that the class results did not come from a process where $\pi = .25$, in other words, that people have a tendency to pick the right front tire more than ¼ of the time?

(f) Use Minitab or the applet to obtain the exact binomial probability. How does it compare to your empirical p-value?

Investigation 3-11: Do Pets Look Like Their Owners? (cont.)

In the actual study conducted by Roy and Christenfeld (2004), they took pictures of 45 dogs and their owners at several local parks in San Diego, CA. For each owner, student judges were shown a picture of the owner, a picture of the owner's dog, and a picture of another dog not belonging to this owner. Each judge was asked to specify which dog they believed was the owner's. If more than half of these twenty-eight student judges correctly paired a given dog with his or her owner, then that owner was considered to be successfully "matched" to his/her dog. This process was repeated for each of the 45 owners.

(a) Suppose the judges are simply guessing between the two dogs. Let X count the number of judges that correctly match the owner with his or her dog. Can X be reasonably modeled by a binomial distribution? Justify each criterion of the binomial random variable and state the values of n and π.

(b) If all twenty-eight judges are simply guessing between the two dogs, use the binomial distribution to determine the probability of a "match" for one particular owner.

(c) Let the random variable Y count the number of successful matches among the 45 owners. Can Y be reasonably modeled by a binomial distribution? Justify each criterion of the binomial random variable and state the values of n and π.

(d) For the binomial distribution in (c), what is the expected number of matches among these 45 owners?

(e) Roy and Christenfeld found that the judges matched the correct dog to 23 of the 45 owners. If all of the judges are simply guessing between the two dogs, is this a surprising result? [*Hint*: Define the parameter in words, state the null and alternative hypotheses, and then use the binomial distribution for Y to calculate the p-value of at least 23 matches in the 45 trials.]

Parameter:

H_0: H_a:

p-value:

Reject or fail to reject H_0:

Conclusion (in context):

(f) The researchers suspected that perhaps owners of *pure-breed* dogs would be matched with their owners more than chance would predict. Twenty-five of the dogs in the study were pure-breeds and sixteen of the pure-breeds were correctly matched with their owners. If the judges are simply guessing between the two dogs, is this a surprising result? Repeat the above test procedure to address this question.

Study Conclusions: The focus in this study is π, the probability of a correct match in a dog-owner pair. We are assuming that the value of π is the same for every dog-owner pair and that if one dog-owner pair is correctly matched, this does not change the probability that the next dog-owner pair is correctly matched. Thus X, the number of correct matches for these 45 dogs, can be modeled by the binomial probability distribution. If the judges are simply guessing each time, the probability of a correct match is P(correct match) = P(Y>14) where Y is binomial with $n=28$ and $\pi=.5$. This probability equals 1-.5747 = .4253. Therefore, our p-value is the probability that X is at least 23 when $n=45$ and $\pi = .4253$, which is .1554. This p-value is too large to provide any evidence that judges do better than just guessing. However, if we look at the pure-breeds only, our p-value is P(X≥16) when $n=25$ and $\pi=.4253$, which is .025. This fairly small p-value is statistically significant at the .05 level. In conclusion, we do not have convincing evidence that the judges are doing better than guessing which dog belongs to each owner, except possibly for the pure-breed dogs. (However, if we apply Fisher's Exact Test to compare these two groups, the p-value is .051, a marginally statistically significant difference between the pure-breed dogs and the "mongrels.")

Investigation 3-12: Halloween Treat Choices

Obesity has become a widespread health concern, especially in children. Researchers believe that giving children easy access to food increases their likelihood of eating extra calories. A recent study (Schwartz, Chen, and Brownell, 2003) examined whether children would be willing to take a small toy instead of candy at Halloween. They had seven homes in 5 different towns in Connecticut present children with a plate of 4 toys (stretch pumpkin men, large glow-in-the-dark insects, Halloween theme stickers, and Halloween theme pencils) and a plate of 4 different name brand candies (lollipops, fruit-flavored chewy candies, fruit-flavored crunchy wafers, and "sweet and tart" hard candies) to see whether children were more likely to choose the candy or the toy. The houses alternated whether the toys were on the left or on the right. Data were recorded for 284 children between the ages of 3 and 14 (who did not ask for both types of treats).

(a) Identify the observational units and variable of interest in this study. Is this variable quantitative or categorical? If categorical, how many different outcomes are possible?

Observational units:

Variable of interest: Type:

Possible outcomes:

(b) Define (in words) the parameter of interest in this study.

(c) If there is no real preference for either type of item, what is the probability that a child will chose the toy item? Is this a null or an alternative hypothesis?

(d) If the null hypothesis is true, how many of the 284 children would you expect to choose the toy?

In this study, the researchers wanted to know if there was a preference for *either* item. They did not have a prior suspicion or a conjectured direction about which item would be preferred. In such cases, we will use a *two-sided* or "not equal to" alternative, H_a: $\pi \neq .5$.

(e) In the sample of 284 children, 135 children chose the toy. Is this more or fewer children than we would have expected if there was no preference?

(f) Use the "Simulating Binomial Distribution" applet to determine the probability of this many or fewer children choosing the toy (under the assumption that the toy and candy are equally preferred).

When stating a two-sided alternative hypothesis, we have to consider outcomes "as extreme or more extreme" in the other direction as well.

(g) What is the difference between the observed number of children and the expected number of children?

(h) What is the probability that at least 7 more children than expected would have chosen the toy? [*Hint*: This is equivalent to the probability that *at most* 135 children would choose the *candy*.]

Def: The *two-sided p-value* is the probability of observing a sample result at least as extreme as the observed result *in either direction*.

So we could find the probability that at most 135 children choose the toy *or* at least 149 children choose the toy (and at most 135 children choose the candy), $P(X \leq 135 \text{ or } X \geq 149) = P(X \leq 135) + P(X \geq 149)$. We can add these probabilities since the two events are not overlapping. Because of the symmetry of the binomial distribution when $\pi = .5$, these two probabilities are the same, so adding them is equivalent to doubling either one of them.

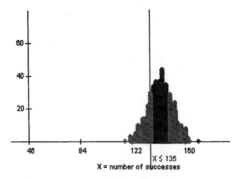

(i) The applet sums these probabilities if you press the "Two-sided" button. Report the two-sided p-value in this study. Are these data significant at the .05 level of significance? Explain.

Study Conclusions: If we define π to be the probability that, when presented with a choice of candy or a toy, a child chooses the toy, and if we assume the null hypothesis (H_0: $\pi = .5$) is true, the above calculations tell us that we would observe at most 135 children choosing candy or at most 135 choosing the toy in about 44% of samples. Thus, this is not a surprising outcome when $\pi = .5$. We fail to reject the null hypothesis in favor of H_a: $\pi \neq .5$. Our conclusion is that children do not seem to have a strong tendency for one type of treat over the other.

We do have some cautions with this study as it was conducted in only a few households in Connecticut, a convenience sample, so we cannot claim that these results are representative of children in other neighborhoods. We also don't know if the children found the toys "novel" and whether their preference for toys could decrease as the novelty wears off. Furthermore, when the children approached the door they were asked their age and gender, and for a description of their Halloween costume. The researchers caution that that this may have cued the children that their behavior was being observed (even though their responses were recorded by another research member who was out of sight) or that they should behave a certain way. Still, these researchers were optimistic that alternatives could be presented to children, even at Halloween, to lessen the exposure to large amounts of candy.

Investigation 3-13: Kissing the Right Way

Most people are right-handed and even the right eye is dominant for most people. Molecular biologists have suggested that late-stage human embryos tend to turn their heads to the right. German bio-psychologist Onur Güntürkün conjectured that this tendency to turn to the right manifests itself in other ways as well, so he studied kissing couples to see if both people tended to lean to their right more often than to their left. He and his researchers observed couples from age 13 to 70 in public places such as airports, train stations, beaches, and parks in the United States, Germany, and Turkey. They were careful not to include couples who were holding objects such as luggage that might have affected which direction they turned. In total, 124 kissing pairs were observed (*Nature*, 2003).

(a) Identify the observational units and the population in this study. Explain how the 124 pairs could be considered a random sample from a process.

(b) Let X denote number of kissing partners who turn to the right in a sample of 124 pairs. Is X a binomial random variable? What assumptions do you have to make? Describe the parameters in this context. Can you specify the values of the parameters?

(c) Let π denote the probability that a kissing couple that is selected at random (from a public place in the United States, Germany and Turkey) turns to the right. What would you suspect is a reasonable conjecture for the value of π? What would your conjecture signify about people's kissing tendencies? If this conjecture is incorrect, do you think π is larger or smaller than this value? Express your answers as a null hypothesis and an alternative hypothesis.

H_0:

H_a:

One possible conjecture is $\pi = .5$, that kissing couples are equally likely to turn right or left. We can consider this as the null hypothesis. The researcher actually suspected that people would turn to the right more often (based on the fact that most people are "right sided") so his alternative hypothesis is that $\pi > .5$. Note that you should always state your hypotheses before you even consider the sample result.

In the actual study, 80 of the 124 couples turned to the right.

(d) Use the "Simulating Binomial Distribution" applet to determine the one-sided p-value for these hypotheses. Is this convincing evidence at the 5% level that kissing couples are more likely to turn to the right? [State the hypotheses in symbols and in words, calculate the p-value, decide to reject or fail to reject the null hypothesis with $\alpha = .05$, and state your conclusion in context.]

H_0:

H_a:

p-value:

Reject or fail to reject H_0?

Conclusion in context:

(e) Dr. Güntürkün noted that about 2/3 of people have a dominant right foot, or eye, and conjectured that people would exhibit a similar tendency of "right sidedness" when kissing. Use the "Simulating Binomial Distribution" applet to determine whether there is statistically significant evidence that the probability of a kissing couple turning to the right *differs* from 2/3. [*Hint*: Use the two-sided p-value.] Adjust your analysis in (d) to reflect this conjecture.

H_0: H_a:

Two-sided p-value:

Reject or fail to reject H_0?

Conclusion in context:

Calculation Details: In calculating a two-sided p-value in the previous investigation, we considered the values that were as extreme as 80 "or more extreme" which we interpreted to be "as far from the expected value." When $\pi = 2/3$, the expected number of successes out of 124 is 82.67 so that the observed value of 80 successes in the sample is 2.67 below this expected value. We can also find $P(X \leq 80)$ to be .336. To find the tail probability in the other direction, you can first determine the values that are as extreme in the other direction, $82.67 + 2.67 = 85.33$, and then you would calculate $P(X \geq 85.33) = P(X \geq 86) = .298$. So then the two-sided p-value reported by the applet was $P(X \leq 80) + P(X \geq 86) = .336 + .298 = .634$. We could also look at the distance in terms of "number of standard deviations."

Another way to interpret "or more extreme" is to consider the outcomes that are even less likely to occur than the value observed. That is, we will include an observed value of x in our p-value calculation if the probability for that value is smaller than the probability of our observed value. In some situations, this will lead to slightly different results. Below is a portion of the binomial distribution with $\pi = 2/3$:

x	79	80	81	82	83	84	85	86	87
$P(X=x)$	.0581	.0655	.0712	.0748	.0759	.0742	.0699	.0635	.0556

Here, $P(X = 80) = .0655$, $P(X = 85) = .0699$, $P(X = 86) = .0635$ so that 86 is the first value of x that has a smaller probability than 80. This alternative method then tells us to report the two-sided p-value as $P(X \leq 80) + P(X \geq 86)$. This leads to the same p-value calculation as above, but there are many situations where it will not. When the sample size is large and the probability of success is near .5, these two methods should lead to very similar results. However, this latter approach (called the *method of small p-values*) accounts for the nonsymmetric shape that the binomial distribution often has with small n and π away from .5. This is the method used by the applet. For example, suppose we had a binomial distribution with $n=20$ and $\pi = .75$, part of which is shown below.

x	10	11	12	13	14	15	16	17	18	19
$P(X=x)$	.0099	.0271	.0609	.1124	.1686	.2023	.1897	.1339	.0669	.0211

Then $E(X) = 15$. If we observe 19 successes, then the symmetric p-value would use $P(X \leq 11)$ but the method of small p-values would use $P(X \leq 10)$.

While we will not advocate one approach over the other here, you will often want to be aware of the algorithm used by your statistical package. In fact, Minitab uses a third approach altogether. Some other software packages find the (smaller) one-sided p-value and simply double it. We find this approach less satisfying when the binomial distribution is not symmetric.

Study Conclusions: These sample data provide strong evidence that there is a tendency for couples to turn to the right more than half the time (one-sided p-value .001) but we are not able to reject the hypothesis that the probability of turning to the right in this kissing process is equal to 2/3 (two-sided p-value = .634). It is quite plausible that this sample result (80 successes out of 124 trials) came from a process where the probability of success was 2/3.

This was not an experiment so we can't draw any cause and effect conclusions as to why couples tend to turn to the right. We should also be cautious in generalizing these results as this was not a true random sample. Based on the study description, the researcher did attempt to get a broad representation of couples, but there could be some hidden sources of bias.

Practice:

3-19) Halloween Treat Choices (cont.)
Reconsider Investigation 3-12, and continue to let π represent the probability that a child would choose the toy.
(a) Suppose we had used the following hypotheses: H_0: $\pi = .5$ vs. H_a: $\pi < .5$. What is this p-value for this set of hypotheses? How does this p-value compare to the two-sided p-value from the above investigation.
(b) Suppose we had used the following hypotheses: H_0: $\pi = .5$ vs. H_a: $\pi > .5$. What is this p-value for this set of hypotheses? How does this p-value compare to the one-sided p-value from part (a)?

3-20) Armrest Battles
In a study reported in 1982 by Hai, Khairullah and Coulmas, researchers observed 426 pairs of passengers in "mixed-sex" seating arrangements on airplanes to see if either person used the joint armrest. Observations were made after a beverage or a meal was served. Passengers who were asleep or obviously couples were not counted. Data were collected by noting whether the man, the woman, both, or neither was using the armrest. Of the 426 pairs, the man used the armrest 284 times, the woman 57 times, both 37 times, and neither 48 times. We will consider only the 341 pairs for which one person or the other used the armrest.
(a) State the hypotheses for testing whether men are more likely than women to use the shared armrest. [Be sure to interpret what π represents in this context.]
(b) Use the binomial distribution to calculate the p-value of this test.
(c) Identify a potential confounding variable to explain the higher armrest use by males in this study.

3-21) Kissing the Right Way (cont.)
Reconsider Investigation 3-13 and continue to let π represent the probability that a kissing couple turns to the right.
(a) Use the applet to determine the two-sided p-value against the alternative hypothesis that π differs from .5. [*Hint*: Use ≥ 80]
(b) Is your p-value in (a) convincing evidence that π differs from .5?
(c) Repeat (a) and (b) for $\pi = .75$. [Use ≤ 80.]
(d) Repeat (a) and (b) for $\pi = .70$.

Investigation 3-14: Kissing the Right Way (cont.)

In the previous investigation, you learned how to decide whether a hypothesized value of the population parameter was plausible based on a one-sided or a two-sided p-value. The two-sided p-value was used when you did not have a prior suspicion or interest in whether the hypothesized value was too large or too small. In fact, in many studies we may not even really have a hypothesized value, but are more interested in using the sample data to estimate the value of the population parameter.

(a) If you observed $\hat{p} = 80/124$ couples turn to the right, what is your best guess for the value of π?

(b) Do you believe π is exactly equal to the value specified in (a)? Do you think it is close? Explain.

We can employ a "trial-and-error" type of approach to determine which values of π appear plausible based on what we observed in the sample. This involves testing different values of π and seeing whether or not the corresponding two-sided p-value is larger than .05. That is, we will consider a value plausible for π if it does not make our sample result look surprising.

(c) You found in the previous investigation and practice problem that .5 and .75 do not appear to be plausible values for π, but .67 and .70 do because the two-sided p-values are larger than .05. Determine the values of π such that observing 80 of 124 successes or a result more extreme occurs in at least 5% of samples. [*Hints*: Use values of π that are multiples of .01 until you can find the boundaries where the two-sided p-values change from below .05 to above .05.]

Def: A *confidence interval* specifies the plausible values of the parameter based on the sample result.

Confidence intervals are an additional or alternative step to a significance test which tells us whether or not we have strong evidence against one particular value for the parameter. Again, statistical software packages employ slightly different algorithms for finding the binomial confidence interval. This investigation should help you understand how to interpret the resulting confidence interval given by the software. What you found in (c) will be called a 95% confidence level since it was derived using the 5% level of significance.

(d) If you were to repeat (c) but using .01 rather than .05 as the criterion for plausibility, would this "99% confidence interval" include more or fewer values than the one based on the .05 criterion? Explain your reasoning.

Minitab Detour: You can use Minitab to perform these p-value and confidence interval calculations for you. Choose Stat > Basic Statistics > 1 Proportion and choose the "Summarized data" option. Specify *n* as the "number of trials" (124 here) and the observed number of successes as the "number of events" (80 here). Then if you click the Options button, you can change the "confidence level", the "test proportion" and the direction. To obtain a confidence interval, the direction needs to be set to "not equal to."

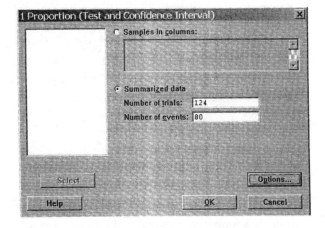

(d) Use Minitab to verify the confidence interval endpoints that you found in (c).

(e) Use Minitab to verify the exact two-sided p-value that you found for testing $\pi = .667$ against the alternative hypothesis that $\pi \neq .667$ based on observing 80 right turning couples out of 124.

(f) Use Minitab to verify the exact one-sided p-value that you found for testing $\pi = .5$ against the alternative hypothesis that $\pi > .5$.

Study Conclusions:

The researchers are assuming they have a representative sample from a process and want to determine π, the probability that a kissing couple turns to the right. Based on a random sample of 124 observations, we estimate π to be close to $80/124 = .645$. However, we know there is some sampling variability, so we want to find an interval of values that appear to be plausible values of π. We do this by finding the values of π for which the two-sided p-value is greater than .05. These are all the values of the population parameter such that our sample result is not overly surprising. You should have found this "95% confidence" interval to be approximately .557 to .726 (results from using Minitab or the applet will differ slightly). Thus, based on these sample results, we are "confident" that the actual value of π is between .56 and .73, i.e., that between 56% and 73% of all kissing couples turn to the right. A 99% confidence interval for π extends from .53 and .75 and therefore includes more values than a 95% interval. The higher level of confidence requires more "room for error." You will learn more about confidence intervals in the next chapter.

Discussion:
In this activity you have learned a second type of "statistical inference" – based on the sample statistic, providing a range of plausible values for the population parameter. Confidence intervals provide a nice companion to tests of significance and are also very useful by themselves. While a test of significance allows to you test a specific hypothesized value, if you reject the null hypothesis, the test of significance provides no information as to *how different* the actual parameter is from the hypothesized value. The confidence interval provides an estimate (with bounds) of the actual value of the parameter.

In fact, there is a type of *duality* between confidence intervals and tests of significance. The confidence interval is the set of values for which we would fail to reject the null hypothesis in favor of the *two-sided* alternative. In fact, Minitab's algorithm for the two-sided p-value is obtained by maintaining this correspondence (the small p-values approach may lead to small departures from this duality). So we can interpret the confidence interval as the set of plausible values for the parameter in that they are the values such that our observed sample result would not be surprising.

In summarizing your results, remember to always conclude with an answer to the research question in context. The decision to reject or fail to reject H_0 should never be your last statement. Similarly, the determination of a confidence interval should never be the last statement. Include at least one more sentence interpreting the results in the context of the problem, e.g., what is supposed to be in the confidence interval and how confident are you that it is!

Summary of Exact Binomial Inference

Let X represent the number of successes in the sample and π the probability of success for the process. If X satisfies the criterion for a binomial random variable (two outcomes, fixed probability of success, independent trials):

To test H_0: $\pi = \pi_0$

We can calculate a p-value based on the binomial distribution with parameters n and π_0. The p-value can be one-sided or two-sided based on the statement of the research conjecture.

> if H_a: $\pi > \pi_0$: p-value = $P(X \geq observed)$
> if H_a: $\pi < \pi_0$: p-value = $P(X \leq observed)$
> if H_a: $\pi \neq \pi_0$: p-value = sum of both tail probabilities using method like "small p-values."

C% Confidence interval for π

The set of values such that the two-sided p-value based on the observed count is larger than the $(1-C)/2$ level of significance.

In Minitab: Choose Stat > Basic Statistics > 1 Proportion.

If the data are presented as a proportion, first convert to the number of successes in the sample, rounding to the nearest integer.

Practice:

3-22) Armrest Battles (cont.)
Reconsider the armrest study, in which 284 of 341 shared armrests were used by the man.
(a) Use a similar procedure as in Investigation 3-14 to construct an interval of plausible values for π, the probability that the shared armrest is used by the male (at the 5% level of significance). Include a one-sentence summary of this interval.
(b) Is .5 among these values? Explain why this is consistent with your earlier analysis.

3-23) Water Oxygen Levels (cont.)
Reconsider the water quality study from Activities 3-9. Recall that in a sample of 34 days, 19 had noncompliant measurements.
(a) Use Minitab to construct an interval of plausible values for the probability of noncompliance for this river based on a 1% level of significance. Include a one-sentence summary of this interval.
(b) Is .1 among these values? Explain why this result is consistent with your earlier analysis.

3-24) Flat Tire (cont.)
Use your class results to obtain a 95% confidence interval for the probability of picking the right front tire. Include a one-sentence summary of this interval.

Practice:

3-25) Type I and Type II Errors
Identify what Type I and Type II Errors would represent for the shopping comparison study and for the armrest study.

3-26) Water Oxygen Levels (cont.)
Recall the dissolved oxygen study with 34 observations (Investigation 3-9). The investigators decided that they wanted to be able to detect a nonattainment rate of 25% and that they wanted the Type I and Type II error rates, considering this alternative value of 25%, to be reasonably similar.
(a) Identify what decisions would be represented by a Type I Error and by a Type II Error in this context. Describe possible consequences from each type of error.
(b) Suppose the quality assessment manager states that the probability of a Type I Error can be at most .15. What is the "cut-off" value for the rejection region? That is, find the smallest x so $P(X \geq x) \leq .15$ when $\pi = .10$. (You can use the applet or Minitab, but clearly describe whether you have found approximate or exact values.)
(c) Was the observed sample result (19 out of 34) in this rejection region?
(d) What is the probability of a Type II Error rate for the cut-off value in (b) and the alternative value of $\pi = .25$ described above? Does this false negative rate appear reasonable, considering the investigators' view that the two types of errors are equally serious?
(e) How would the error rates in (b) and (d) change if instead we made the cut-off 5 or more? Explain intuitively, and then confirm your answer with appropriate calculations.
(f) Would this cut-off suggested in (e) be appropriate if the investigators considered Type I Error to be more serious than Type II Error, or vice versa? Explain.

SECTION 3-4: SAMPLING FROM A POPULATION II

Previously, we applied the binomial distribution to samples from a process (probability of guessing correctly, probability of turning to the right). In these cases, we felt comfortable assuming that the probability of success was constant and that the observations were independent. In many applications, the sample comes instead from a finite population such as the Gettysburg Address example or a Gallup public opinion poll of American adults. In these cases the parameter of interest is the proportion of "successes" in the population, again denoted by π. In this section you will explore the applicability of the hypergeometric distribution and the binomial distribution for such finite populations.

Investigation 3-16: Sampling Words (cont.) (Probability Detour)

Recall from Investigation 3-1 that you selected a sample words from the population of 268 words in the Gettysburg Address and that there were 99 "long" (more than 4 letters) words in this population ($\pi = .369$). In Investigation 3-4, you used the hypergeometric distribution to calculate that the probability of obtaining 5 long words in a random sample of 5 words was .0064.

(a) If we randomly sample *one word* from this population, what is the probability that the word we select is "long"?

If the proportion of successes in the population equals π and we draw one observational unit at random from that population, then the probability of success for that observation is also π.

(b) Suppose after looking at the first word, you put it back and then draw the second word from all 268 words in the population. Would the probability of success be the same as in (a)? Would the trials be independent?

(c) Suppose you randomly select the word "dedicated" and do not replace it into the population. How many "long words" remain? How many "short words"?

    # long: # short:

(d) If a word is randomly selected from the remaining 267 words, now what is the probability that it is a "long word"? Has it changed much?

    P(2nd word is also long):

    Close to previous probability?

> Just as you calculated *conditional proportions* in Chapter 1, here you have calculated a *conditional probability*, the probability of selecting a long word as the second word, knowing that you have already selected a long word as the first word. The notation for this is P(second word is long | first word is long) where the vertical line is read as "given."
> If the two outcomes "first word long" and "second word long" are *independent*, then P(second word is long | first word is long) = P(second word is long | first word is short) = P(second word is long). In other words, the probability of selecting a long word is the same regardless of the length of the first word.

Discussion: If we were to sample with replacement, then the trials would be completely independent and we could apply the binomial distribution. But if you sample without replacement, then the trials are not independent and the binomial criterion would not be met (and you would continue to use the hypergeometric probability distribution). However, in some situations the trials are *almost* independent.

(e) Suppose we randomly select 4 words without replacement. If all of them are long, what is the conditional probability that the 5th word will also be long?

(f) Is the probability of success (long word) on the 5th word much different from the probability of success on the first word even in this extreme case?

(g) Now suppose you have selected 50 consecutive long words (without replacement). Suppose you draw another word. What is the conditional probability that the 51st is also long? Is the probability of success on the 51st word similar to the probability of success on the first word?

Discussion: You will notice that the conditional probability of "success" is now quite a bit lower than in (a). Note that when the sample was small (e.g., $n=2$, $n=5$), the difference in probabilities was not large, but when the sample is a large fraction of the population (e.g., 50 out of 268), then the discrepancy is more noticeable.

In many situations, even when we sample *without replacement*, the conditional probabilities of our subsequent draws do not change dramatically. In this situation, we can assume the trials are roughly independent. As a general guideline, we will consider this approximation valid if the size of the population is at least 20 times the size of the sample. When this is true, the hypergeometric distribution can be approximated by the binomial distribution with n=sample size and π= proportion of successes in the population.

(h) In Investigation 3-4, you took 5 words from the 268 words in the population. Is the binomial approximation to the hypergeometric distribution valid here? What are the values of n and π?

(i) Using the binomial distribution, approximate the probability of obtaining 5 long words in a sample of n=5 words and $\pi = .369$. How does this approximate probability compare to the probability you found using the hypergeometric distribution?

(j) If we take a sample of 10 words, should the binomial approximation still be valid? Explain.

(k) Let X= number of "long words" in the sample of 10 words. Use Minitab to store this complete (exact) probability distribution of X using the hypergeometric distribution in columns 1 and 2. [*Hint*: First put the integers 0 through 10 in one column.] Then use Minitab to display the binomial approximation in a separate column. Are these probability distributions similar? Explain.

(l) Repeat (k) with a random sample of 50 words. Are the probability distributions more or less similar than in (k) where the sample size was 10? Explain.

Summary: These calculations should convince you that the hypergeometric and binomial probabilities are quite similar for *n*=10 but do show some discrepancies when *n*=50. When *n*=50, we do not satisfy the guideline that the population is at least 20 times the size of the sample and the approximation will not be considered valid.

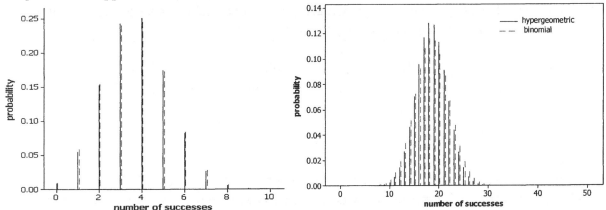

For example, when *n*=50, if we were to calculate P($X \leq 15$) we would get .167 with the hypergeometric distribution but .194 with the binomial approximation.

While it is no longer the case with the omnipresence of computers, historically it was easier to determine probabilities from the binomial distribution (using tables of values) than from the hypergeometric. When we're dealing with populations in the hundreds of millions (such as U.S. citizens), the binomial approximation to the hypergeometric is very accurate and much easier to work with. In fact, in many cases we may not know the exact population size but we can still work with the probability of success in the population and the binomial approximation is very useful. Many of the techniques we will learn in this class still make use of this approximation. Just keep in mind that the approximation only applies when the population size is much, much larger than the sample size.

An interesting consequence of this approximation is that the only parameters are *n* and π. The population size does not affect the binomial probabilities. This parallels what you saw in practice problems 3-9 and 3-10 – if the population is much larger than the size of the sample, changing the size of the population does not appreciably affect the behavior of the sampling distribution.

Binomial approximation to the hypergeometric: If the parameter of interest is the proportion of successes in a population (π) and we take a random sample from that population, then when the population size is more than 20 times the sample size we can apply the binomial distribution (*n*, π) instead of the hypergeometric distribution for the number of successes in the sample.

You should note that we have calculated probabilities in terms of $\hat{p}$ and in terms of *X* interchangeably, e.g., P(*X*=5) = P($\hat{p}$ =1) when *n*=5. Thus, this approximation guideline tells us when we can use the binomial distribution as a description of the sampling distribution of sample proportion.

Investigation 3-17: Feeling Good

The Harris Poll conducts a yearly survey on whether or not people "feel good" about various aspects of their lives. In 2003, the poll was conducted on a nationwide cross-section of 1017 adults on October 14-19. They used techniques slightly more involved than "simple random sampling" to help ensure their sample is representative of the population.

(a) Do these 1017 adults constitute a sample or a population? Explain.

(b) Identify the population of interest for this study.

(c) One of the questions participants were asked about was "the quality of your life overall." Take a guess for the proportion of all adult Americans who would answer that they feel good about the quality of their life overall. Is this a guess for the value of a parameter or a statistic?

(d) Using your guess for (c), what is the probability that a randomly selected adult will say that he/she feels good about their overall quality of life?

(e) Is the binomial approximation to the hypergeometric distribution valid here? Explain. [Assume that Harris used a simple random sample.]

(f) In this Harris poll, 88% (895) of respondents said they felt good about their overall quality of life. Carry out a test to decide if there is strong evidence against your guess for π at the .05 level. [*Hint:* Convert .88 to the nearest integer number of successes. If the approximation is valid, find the two-sided p-value using Calc > Basic Statistics > 1 Proportion, otherwise use the hypergeometric distribution.] State the hypotheses, p-value, and your conclusion.

   H_0: H_a:

   p-value: Conclusion:

(g) Explain what a Type I and a Type II error represent in this context. Which type of error might you be committing? Explain.

Type I Error:

Type II Error:

Which here?

(h) Which values tested in your class were not rejected at the .05 level? Which were rejected at the .05 level?

(i) Using the binomial approximation, report Minitab's 95% confidence interval for the population proportion based on this sample. Include a one-sentence summary of this interval. Does this interval include your conjectured value? Explain how you already knew the answer to this question from your answer to (f). Does the interval appear to correspond to the values mentioned in (h)? Explain.

Study Conclusions: Since we have a random sample and the population is much larger than the sample, we can apply the binomial approximation to the hypergeometric distribution. We will consider "success" to be that the respondent answers that they feel good about the quality of their life overall. Then our binomial parameters are $n=1018$ and π, the proportion of all adult Americans who would respond that they feel good about the quality of their life overall. A 95% binomial confidence interval for π is (.858, 899), indicating the plausible values for π based on our sample. Any value not in this interval is rejected by a two-sided test of significance.

Practice:

3-27) Committee Members (cont.)
Recall from Investigation 1-10 that 2 people were selected to represent a committee of 6 faculty members, and we wanted to know whether it was surprising for both women to be selected.
(a) Define the random variable for this context.
(b) Suppose one person is randomly selected for the subcommittee. Calculate the probability that this person is female.
(c) Suppose a second person is then randomly selected for the subcommittee. Calculate and compare the following conditional probabilities:

P(second person selected is female| first person selected is female)
P(second person selected is female| first person selected is male)

(d) Suppose one person has been selected but we don't know whether this person is male or female. To calculate the probability that the second person is female we consider the two possibilities: P(second person female) = P(both female *or* first male and second female)
= P(FF) + P(MF)
= P(second female|first female)P(first female) + P(second female|first male)P(first male)
Explain each of these calculation steps using the probability rules discussed in Investigation 3-8 and finish the calculation. How does your result compare to (b)?
(e) Suppose we randomly select two people from this committee. Is P(female) constant for each person selected? Are the outcomes for each individual independent? Explain.

3-28) Feeling Good (cont.)
(a) Recent Census results indicate that there are roughly 209,128,094 adult Americans. Using your guess in (c) of Investigation 3-15, how many adult Americans feel good about the quality of their life overall? How many do not feel good?
(b) Suppose that the first 1016 adults you interview all say that they feel good about their quality of life and you plan to randomly select one more adult. What is the probability that this next person will also say they feel good about their quality of life?
(c) How does your calculation in (b) compare to your guess in (c)? Has this probability changed much? Does it seem reasonable to consider this probability constant for this sample?
(d) Repeat (f) in Investigation 3-17 using the hypergeometric distribution. How does the probability compare to that calculated with the binomial approximation?

3-29) Popcorn Workers (cont.)
Recall the popcorn workers study from Investigation 1-1. Another issue is how these workers compared to the general population. Information about the general population was obtained from the National Health and Nutrition Examination Survey (NHANES III). The NHANES study found 8% of current or former smokers with airway obstruction, 2.3% of nonsmokers with airway obstruction and 5.5% as the baseline rate of airway obstruction in the general population. At the popcorn-plant, they observed 8 of 64, 13 of 52, and 21 of 116 workers with airway obstruction in the smoker, nonsmoker, and overall groups, respectively. Of interest is whether the rate of airway obstruction is higher among the population of popcorn workers compared to the national results.
(a) Consider the population of all microwave popcorn workers who are current or former smokers. Let π represent the proportion of this population with airway obstruction. Carry out a

test of significance to determine whether there is evidence that the airway obstruction rate in this population is higher than national rate for former and current smokers. [State your hypotheses (in symbols and in words), determine whether you can apply the binomial approximation in this study, report the (approximate) p-value, and state your conclusion in context.]

(b) Repeat (a) for the population of microwave popcorn workers who are nonsmokers and the corresponding national rate.

(c) Repeat (a) for the population of microwave popcorn workers and the corresponding national rate.

(d) Which group (or groups) has the strongest evidence that the rate of airway obstruction among popcorn factory workers is higher than the national rate? Explain.

Investigation 3-18: Long-term Effects of Agent Orange

Numerous studies have examined the presence of 2,3,7,8-TCDD – the dioxin contaminant of Agent Orange – in both the food sources and blood levels of Vietnamese citizens. Recent studies (e.g., Schecter et al, 2003) looked at the blood TCDD from residents of Bien Hoa City. Bien Hoa City was the location of an airfield used for Agent Orange storage and spray missions during the war. There were several spills at the airfield, especially in 1970. The authors state that non-exposed persons average 2 parts per trillion (ppt) and elevated TCDD is defined as >5 ppt. A sample of 43 residents found that 41 had elevated levels of TCDD.

(a) Is this an experiment or an observational study? Explain.

(b) Define the population of interest in this study.

(c) Do you know the size of the population in (b)?

(d) Let X represent the number of residents with elevated TCDD levels in a random sample of 43 residents. Do you think the binomial approximation to the hypergeometric distribution is valid here? Explain.

(e) Carry out a test of significance to determine if there is convincing evidence that more than half of all current residents in Bien Hoa City have elevated levels of TCDD. [State the hypotheses, p-value, and conclusion in context.]

(f) Suppose you had failed to reject the null hypothesis and therefore concluded $\pi = .5$. What would this conclusion imply about the *median* TCDD level in this population?

In the above test, the underlying variable (TCDD level) was quantitative. Instead of analyzing this variable, the measurements were converted into a binary variable by counting the number of observations above 5 ppt. It was then possible to apply the binomial distribution. If π represents the proportion of the population above 5 ppt, a test of H_0: $\pi = .5$ is equivalent to a test of H_0: population median = 5 ppt. This type of analysis (testing the population median through the binomial distribution) is often referred to as the *Sign Test*. It is especially useful for *paired* data, where you can count the number of pairs with a larger (or smaller) value.

Study Conclusions: If π represents the proportion of all Bien Hoa City residents with elevated TCDD levels, the above analysis indicates that there is very strong evidence (p-value ≈ 0) against the null hypothesis (H_0: $\pi = .5$) in favor of the alternative hypothesis (H_a: $\pi > .5$).
We conclude that there is strong evidence that the population median TCDD level is larger than 5 ppt, indicating that a majority of this population has elevated TCDD levels. The above analysis assumed that the population of Bien Hoa City residents is larger than 20×43 and that the sample was representative of all current residents. Both of these details would need to be verified before we can consider these conclusions valid.

Practice Problems:

3-30) Long-term Effects of Agent Orange (cont.)
(a) All 43 members of this sample had TCDD levels above 2. Suppose we wanted to test whether the population median was larger than 2 ppt. How would the p-value for this test compare to the p-value in the previous investigation? Explain.
(b) Describe how you could test whether the third quartile in this population was significantly different from 5 ppt.

3-31) Schizophrenic Twins
Scientists have long been interested in whether there are physiological indicators of diseases such as schizophrenia. In a 1990 study by Suddath et. al., reported in Ramsey and Schafer (2002), researchers used magnetic resonance imaging to measure the volumes of various regions of the brain for a sample of 15 monozygotic twins, where one twin was affected by schizophrenia and other not ("unaffected"). The twins were found in a search through the United States and Canada, the ages ranged from 25 to 44, years with 8 male and 7 female pairs. The data (in cubic centimeters) for the left hippocampus region of the brain appear below and in the Minitab worksheet Hippocampus.mtw:

Pair #	Unaffected	Affected	Pair #	Unaffected	Affected	Pair #	Unaffected	Affected
1	1.94	1.27	6	1.66	1.26	11	1.25	1.02
2	1.44	1.63	7	1.75	1.71	12	1.93	1.34
3	1.56	1.47	8	1.77	1.67	13	2.04	2.02
4	1.58	1.39	9	1.78	1.28	14	1.62	1.59
5	2.06	1.93	10	1.92	1.85	15	2.08	1.97

The primary research question is whether the data provide evidence of a difference in hippocampus volumes between those affected by schizophrenia and those unaffected.
(a) What are the observational units in this study? (Be specific.)
(b) Is this an observational study or a controlled experiment? Explain.
(c) Explain why the researchers went to great lengths to find same-sex *twins* to study. Specifically, identify some potential confounding variables that are controlled by studying twins.
(d) Let X represent the number of twins where the unaffected twin had a larger left hippocampus volume. If there was no tendency for either twin to have a smaller left hippocampus volume, would X follow a binomial distribution? Explain. With what value of π?
(e) Create an indicator variable that determines whether or not the unaffected twin had a larger left hippocampus:
     MTB> let c4=(c1-c2>0)   MTB> tally c4
How many unaffected twins had a larger left hippocampus?
(f) Use the binomial distribution (*sign test*) to determine the p-value for obtaining at least this many unaffected twins having the larger left hippocampus, assuming that this is true for half of all monozygotic schizophrenia-discordant twins.
(g) Calculate the two-sided p-value and explain what this quantity measures.
(h) Does this analysis provide statistically significant evidence that the one twin is more likely to have the larger left hippocampus than the other? Explain.
(i) In the discordant twin schizophrenia study, 13 of the unaffected twins had a larger right hippocampus. Would the p-value for the sign test be larger or smaller than that reported above? Explain.

SUMMARY

In this chapter, you switched from comparing groups to considering issues that arise when you draw a sample from a larger population. The first step is to ensure that the sample will be representative of the larger population. This is done by using random mechanisms to select the sample. These random mechanisms (e.g., simple random sampling, stratified sampling, cluster sampling, systematic sampling) lead to unbiased samples that will have characteristics mirroring those of the population. You must also be cautious about avoiding nonsampling errors. Even with random sampling and other precautions, there will be random sampling variability – the characteristics of the sample will vary from sample to sample. Fortunately for all statisticians, this sampling variability follows a predictable pattern, allowing us to make predictions about how far our sample results will typically fall from the population value. When dealing with a binary, categorical variable, the sample result can be modeled as coming from a Bernoulli process or from a very large population. Using the binomial distribution, we can formally calculate probabilities for different sample results. This allows us to again calculate p-values to measure how unlikely we are to get a sample result at least as extreme as what was observed under certain conjectures about the population or process from which the sample was drawn. As before, when the p-value is small, we have evidence that the observed result did not happen solely due to the random chance inherent in the sampling process. The smaller the p-value, the stronger the evidence against the null hypothesis. Still, we must keep in mind that we are merely measuring the strength of evidence - we may be making a mistake. If we decide the result did not happen by chance, there is still a small probability that it did (the probability of committing a Type I Error). If we decide the result did happen by chance, there is still a probability that something other than random chance was involved (a Type II Error). It is important to consider the probabilities of these errors when completing your assessment of a study. In particular, the sample size of the study can influence the probability of a Type II error.

You also began your study of confidence intervals as specifying a range of plausible values for the population parameter based on what you observed in the sample. These were the hypothesized values that generated two-sided p-values above the level of significance α. In other words, they were the values for which your sample result would not be surprising. You will learn more about confidence intervals in later chapters.

In this chapter, we repeated the process of describing the sample data and then considering making inferential statements about the population. In particular, before you can decide whether your sample result is surprising, you need to know what sample proportions typically look like when the null hypothesis is true. You first gained this insight through simulations and then, in the case of categorical data, more formal probability models. You will learn more about probability models, including models for the sampling distribution of the sample *mean*, in the next chapter.

It is important that you keep in mind that which distribution to use (e.g., hypergeometric or binomial) depends on certain "conditions" that need to be checked and verified before proceeding (is the probability of success constant, are the observations independent, is the population very large). Roughly, we will use the hypergeometric distribution when we are sampling from small populations without replacement, we will use the binomial distribution when we are sampling from a process or from a large population.

SUMMARY OF WHAT YOU HAVE LEARNED IN THIS CHAPTER:
- Distinction between sample and population and process
- The term parameter to describe a numerical characteristic of a population or process
- The symbol π to represent a population proportion or the probability of success for a process and μ to represent a population mean.
- Random sampling eliminates bias and allows us to use results from our sample to represent the population. There are other sampling methods if a random sample is not feasible.
- Ways to try to avoid nonsampling errors in your sample survey
- The fundamental notion of sampling variability and how to simulate empirical sampling distributions
- The criteria for a Bernoulli process/binomial distribution
- Using Minitab to calculate binomial probabilities, as well as exact p-values and confidence intervals
- The formal structure of a test of significance about a population parameter (define the population parameter, determine which probability model to use, calculate the p-value as specified by the alternative hypothesis under the assumption that the null hypothesis is true, decide to reject or fail to reject the null hypothesis for the stated level of significance, and state the conclusion in context). You should notice that the interpretation of the p-value (could this result have happened by chance) is consistent with the p-values you calculated for experiments and will be the same throughout this text.
- What factors affect the size of the p-value
- Type I and Type II errors: what they mean, how their probabilities are determined, and how they are affected by sample size and each other
- The idea of a confidence interval as the set of plausible values of the parameter that could have reasonably led to the sample result that was observed.
- The definition of conditional probability
- A criterion for probability events to be independent
- The binomial approximation to the hypergeometric distribution when sampling without replacement or from a very large population.

TECHNOLOGY SUMMARY
- In Minitab, you learned how to calculate exact binomial p-values and confidence intervals by choosing Stat > Basic Statistics > 1 Proportion.
- You used java applets to explore sampling distributions. The "Sampling Words" applet is tied to the Gettysburg Address activity but allows you to explore the sampling distributions of both sample means and sample proportions for that context. The "Simulating Power" applet allowed you to compare the distribution under the null hypothesis to the distribution under an alternative hypothesis and consider the probabilities of Type I and Type II errors and how they are controlled. The "Simulating Binomial Distribution" applet allowed you to explore properties of the binomial distribution. This applet provides both empirical and exact binomial probability calculations

Probability Detour: In the above practice problems you have seen some distinctly non-normal distributions. There are many other probability models that do an excellent job of describing skewed distributions. The *exponential probability model* is used quite often in practice, especially to model interarrival times for processes whose arrivals occur with a constant rate.

The probability curve is given by the function: $f(x) = \frac{1}{\beta}e^{-x/\beta}$, $x \geq 0$, where β is the only

parameter that needs to be specified and represents the average interarrival time. The *gamma probability model* is useful in that it is more flexible and can represent a wide variety of skewed shapes. It is often useful to measure "lifetimes," including customer service times or time of machine repair. The functional form of the probability curve is more complicated and involves two different parameters, one considered a shape parameter (α) and the other considered a scale parameter (β): $f(x) \propto x^{\alpha-1}e^{-x/\beta}$, $x \geq 0$ (to complete the specification of this function you would need to determine the multiplicative constant for which the area under the function is one). Having two parameters is what provides more flexibility in modeling different behavior but at the cost of needing to determine the second parameter. The exponential probability curve is actually a special case of the gamma probability curve.

An Exponential model with $\lambda=2$:

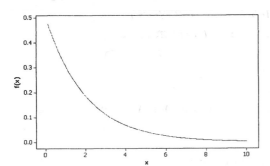

A Gamma model with $\alpha=2$, $\beta=1$:

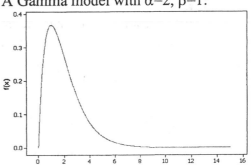

A Gamma model with $\alpha=2$, $\beta=2$:

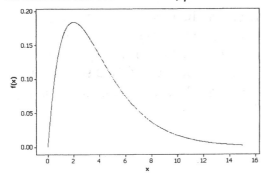

A Gamma model with $\alpha=3$, $\beta=1$:

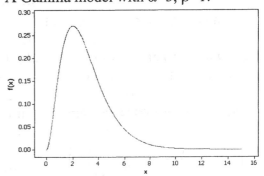

These are just a few of the possible exponential and gamma distributions as well as just a few of the possible models for skewed data.

Notice that these functions are scaled so the total area underneath each curve is one.

Practice:

4-1) Earthquakes
The United States Geological Survey maintains data on all earthquakes in the USA and around the world (http://earthquake.usgs.gov/). The file earthquake.mtw contains data for all earthquakes of magnitude greater than 1 in the US between March 25 and April 1, 2004.
(a) Create a histogram of these earthquake magnitudes (C1) with the normal probability curve "fit" overlaid. Does the normal model appear to do a good job of describing these data? Explain.
(b) Double click on the blue density curve in Minitab to open the Edit Distribution Fit window, select the Options tab, and use the pull-down menu to change the Distribution to Gamma. Does this probability curve appear to be a better model for these data? Explain.
(c) How do you think this distribution of earthquake magnitudes would differ if the earthquakes of magnitude < 1 were also included?

4-2) Aussie Births
The file aussiebirths.mtw contains data on births for 44 babies born in one 24-hour period in Brisbane, Australia (http://www.amstat.org/publications/jse/datasets/babyboom.txt). This was a record high number of births in one day. We want to explore the distribution of "time between births." Note that C4 contains the times of the births (in minutes after midnight).
(a) Use Minitab to calculate the time between births:
```
MTB> diff c4 c5
```
Verify that this "diff" command calculates the time between births (and name the column). Then produce a histogram of the "time between births" variable. Describe the characteristics of this distribution.
(b) Use Minitab to overlay a normal probability curve on this histogram [*Hint*: Right click on the Graph and select Add > Distribution Fit]. Does the normal model do a reasonable job of describing these data?
(c) In Minitab, double click on the blue curve in the histogram window, select the Options tab, and change the Distribution to Exponential. Does this probability curve appear to be a better model for these data? Explain.
(d) In Minitab, to start the first histogram bin at zero (can't have negative times!), double clicking on the horizontal scale to open the Edit Scale window. Then in the "Scale Range" box, uncheck the box and enter 0 as the minimum value. Then press the binning tab and choose Cutpoint instead of Midpoint as the interval type. Does this revised histogram change your answer to (c)?

Practice
4-3) Aussie Births (cont.)
(a) Produce a normal probability plot of the times between births. Describe how the distribution deviates from normality.
(b) Produce and describe an exponential probability plot of the times between births. [You can double click on the blue lines and in the Edit Distribution Fit window, use the pull-down menu to change the distribution from Normal to Exponential. When creating a new graph, you can get to this window under the Distributions button.]
(c) Take a log transformation of the times (`MTB> let c6=log(c5)`). Produce a normal probability plot of these transformed data. Does this plot suggest that a normal model might be appropriate for describing the distribution of the logs of times? Explain.
(d) Take a square root transformation of the times (`MTB> let c7=sqrt(c5)`). Produce a normal probability plot of these transformed data. Does this plot suggest that a normal model might be appropriate for describing the distribution of the square roots of the times? Explain.

4-4) Matching Probability Plots to Boxplots Graphs for three different variables are given below, one boxplot and one normal probability plot for each. Which boxplot corresponds to which normal probability plot? Write a few sentences providing your justification.

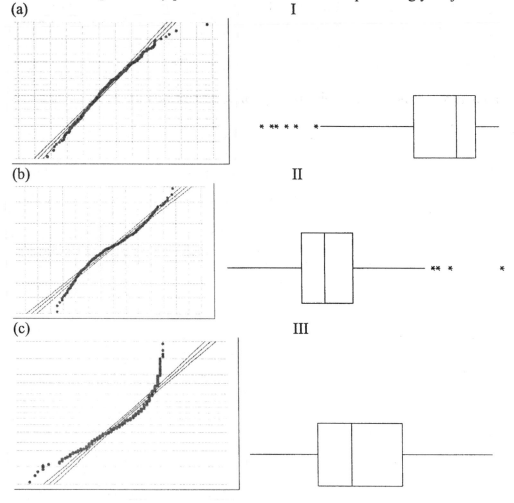

SECTION 4-2: APPLYING THE PROBABILITY MODEL

Now that you know some tools for helping you decide whether or not a variable is well modeled by the normal distribution, you will learn that you can make probability statements about the values the variable will take.

Investigation 4-3: Fuel Capacity

(a) Return to the `misc.mtw` worksheet and re-produce a histogram of the fuel capacities of different car models with a normal probability curve overlaid. Report the mean and standard deviation of these data.

$$\bar{x} = 16.38 \quad \sigma_x = 7.708$$

(b) According to the empirical rule, where should the middle 68% of the fuel capacity values lie?

$$13.68 \le x \le 19.08$$

(c) Use Minitab to determine the percentage of these fuel capacities that fall within one standard deviation of the mean:

```
MTB> let c9=(c8>mean(c8)-std(c8) & c8<mean(c8)+std(c8))
MTB> tally c9
```

Convert this count to a percentage. How well does this percentage match that predicted by the empirical rule?

68.5%   this is off by .5%

> Checking the empirical rule is another way to support whether the data are reasonably modeled by a normal distribution.

(d) Suppose we want to determine the probability of a randomly selected car model having a fuel capacity of 13 gallons or less. From your histogram, estimate the proportion of the distribution that lies below 13.

~ 15%

> If the proportion of values lying below x is p, then if we randomly select one value from the data set, the probability its value is below x is also p, $P(X \le x) = p$. If we assume the same model applies to a larger population of observational units, then this gives us an idea of the probability of obtaining a value less than x from the population.

Often, we don't have a data set from which to determine the proportion of observations falling below a value, but instead we rely on the probability model of the population. To calculate a probability from a density curve, we need to determine the *area under the curve* for the region of interest.

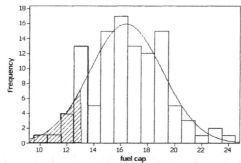

We can do this using geometry in some cases, using integration in others. But the normal probability density function has no closed form expression for its integral, so we will rely on technology (Minitab and a Java applet) to calculate areas/probabilities through numerical integration techniques.

(e) To use the normal probability model to estimate the probability of a car model holding fewer than 13 gallons, you want to find the area under the normal curve to the left of 13. To do this in Minitab, choose Calc > Probability Distributions > Normal. Leave the Cumulative probability button selected, specify 16.38 as the mean and 2.708 as the standard deviation. Click the button next to Input constant and specify 13 in the box. Click OK.

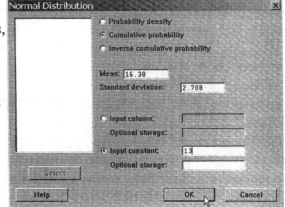

What did Minitab report for the probability of a car model having less than a 13-gallon capacity?

$$P(X \leq 13) = .108937$$

(f) Recalling the relative frequency interpretation of probability, write one sentence explaining what the value calculated in (e) represents.

The prob of a randomly selected car having a fuel capacity below 13 gallons.

(g) For the sample of 108 cars models, determine how often the fuel capacity was less than 13 gallons: MTB> let c10=(c8<=13)
         MTB> tally c10
Convert the count to a proportion and discuss how this value compares to the value predicted by the normal distribution model in (e).

$$\frac{11}{108} = .1019$$ which is very close to the predicted

Discussion: For these data, the normal model appears to do an reasonable job of predicting how these values will behave. In this case, we had data with which to confirm that a normal model was appropriate. Often, we will have to use a little more faith that a model is appropriate, keeping in mind that our predictions are only accurate if the model does indeed adequately summarize the behavior of the variable.

Practice:

4-5) Earthquakes (cont.)
Suppose we model the U.S. earthquake magnitudes with a Gamma distribution with parameters 10.74 (shape) and .1860 (scale). Use Minitab to determine the probability of an earthquake with magnitude greater than 3 using this Gamma model.
(Calc > Probability Distributions > Gamma)

4-6) Triangular Density
Suppose we thought the following density curve is a reasonable model for a data set:

$$f(x) = \begin{cases} x & 0 \le x \le 1 \\ 2-x & 1 < x \le 2 \\ 0 & otherwise \end{cases}$$

(a) Sketch this density function by plotting $f(x)$ vs. x where x ranges from -1 to 3.
(b) Verify that the area under the function is equal to one, both through geometry and integration.
(c) Calculate the probability of $X > 3/2$ by using your sketch (and a little geometry) to determine the area under the density curve above 3/2.
(d) Verify that you could obtain the answer to (c) through integration as well.

4-7) Aussie Births (cont.)
Suppose we think that the times between births in an Australian hospital are well modeled by an Exponential distribution with parameter $\lambda=33$ minutes and you want to determine the probability of more than one hour (60 minutes) transpiring between births. (See p. 4-6.)
(a) Write the function for the density curve with this value of λ.
(b) Integrate this function to determine $P(X \ge 60)$.
(c) Use Minitab to confirm your calculation (scale = 33, threshold = 0).
(d) Use Minitab to determine how many of the 43 observed times between births were longer than 60 minutes. How does this relative frequency compare to the probability predicted by the model?

Investigation 4-4: Body Measurements (cont.)

Recall the body measurements from Investigation 4-2. Another variable recorded by the researchers was the heights of the individuals.

(a) Do you think the heights will follow a normal distribution? Do you think there will be differences in the distributions of heights for males and females? Explain.

Yes and Yes

(b) Open BodyMeasurementHeights.mtw and produce normal probability plots of the heights for men and women in this study. Do these distributions appear to be reasonably modeled by a normal distribution? How do the distributions differ?

Yes they seem reasonably to fit

they have diff centers but roughly the same σ

(c) Would you consider someone with a height of 185 cm surprising? Explain.

Not if they were male.

Yes if female.

Suppose we want to model female heights as Normal with mean $\mu = 164.9$ cm and standard deviation $\sigma = 6.55$ cm and male heights as Normal with mean $\mu = 177.7$ cm and standard deviation 7.18 cm. We will abbreviate these models as N(164.9, 6.55) and N(177.7, 7.18).

(d) Sketch these two models on the horizontal axis below, being clear how the mean and standard deviation are represented in each sketch.

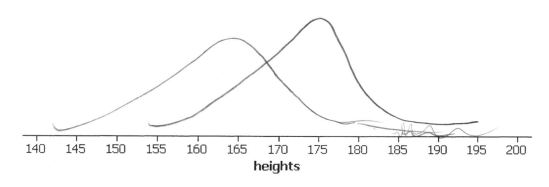

(e) Shade the area under each curve corresponding to a height of at least 185 cm.

(f) We can use Minitab to calculate the probability of randomly selecting a female (from a distribution with mean $\mu = 164.9$ and standard deviation $\sigma = 6.55$) who is at least 185 cm tall. Choose Calc > Probability Distributions > Normal. Make sure the "cumulative probability" option is selected and specify the mean and standard deviation values. Enter 185 in the "Input constant" box. Click OK.

What does Minitab report? $P = .998925$

(g) You will note that Minitab always reports the probability to the left of the input constant, $P(X \leq x)$. What is the total area under the curve and how will the probability above a value be related to the probability below a value? When you have completed the calculation, check your sketch to confirm that your numerical answer appears reasonable.

Total $A = 1$

$P(X \geq 185) = 1 - .998925 = .001075$ appears reasonable

Calculation Hint: With the normal probability model, you can assume $P(X \leq x) = P(X < x)$ since the model must assign $P(X=x)$ to be equal to zero (because the "area" of a line above a single point is equal to zero). You may remember a similar discussion from doing integrals in calculus. This was certainly not true when we were working with a discrete random variable (e.g., binomial) where we had to be very careful whether we intended $P(X > x)$ or $P(X \geq x)$ since $P(X=x)$ had a nonzero value.

(h) Use Minitab to calculate the probability of randomly selecting a male that is at least 185cm tall, assuming male heights follow a N(177.7, 7.18) distribution. How do these probabilities compare?

$P(X \geq 185) = 1 - .884355 = .115645$

Z-SCORES
(i) Using the above model, calculate the *z*-score for a height of 185 cm for a female and then for a male.

z (female): $\dfrac{185 - 164.9}{6.55} = 3.07$ z (male): $\dfrac{185 - 177.7}{7.18} = 1.01671$

How do these *z*-scores compare?

They are very different

Discussion: As you saw in chapter 2, standardizing observations allows us to measure the distance between the observation and the mean so that we can compare across distributions with different centers and spreads. A height of 185 cm is more standard deviations above the mean female height than above the mean male height.

In fact, we can standardize the entire distribution.

(j) Use the `let` command to standardize all of the female heights:

    MTB> let c5=(c3-mean(c3))/std(c3)

Create and describe a histogram of this distribution and determine the mean and standard deviation of this distribution of z-scores.

This histogram seems reasonably normal with a $\mu \approx 0$ & $\sigma = 1$.

(k) Repeat (j) for the male heights.

$\mu \sim 0$
$\sigma = 1$ looks roughly normal

Discussion: You should see that when you standardize each distribution, this is a linear transformation that shifts the distribution to center at zero and scales the distribution to have a standard deviation of 1. That is, standardizing any normal distribution converts it to a N(0,1) distribution.

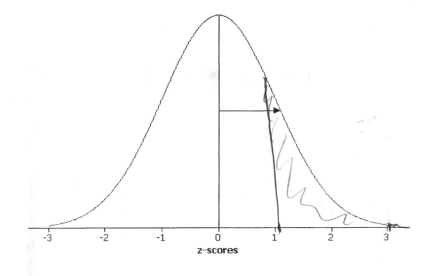

z-scores

Def: The *standard normal distribution* follows the normal probability curve with mean $\mu = 0$ and standard deviation $\sigma = 1$. We will continue to use the letter Z to refer to this distribution.

You should also note that almost all of the observations lead to a z-score between -3 and 3. That is, almost all of the observations in a normal distribution fall within 3 standard deviations of the mean (99.7% according to the empirical rule).

(l) Mark the location of the z-scores you computed in (i) on the graph above to help judge the relative position of each observation in this distribution.

We can convert these z-scores to probabilities as well.

(m) Choose Calc > Probability Distributions > Normal and enter the female z-score. Specify 0 as the mean and 1 as the standard deviation. What is the probability lying above this z-score in the standard normal distribution?

$$P(Z \geq 3.0687) = 1 - .998925 = .001075$$

Note: You can also compute this probability by typing `cdf 3.0687` at the MTB prompt and then subtracting the result from one.

(n) Repeat (m) for the male z-scores. How do these probabilities compare to those you determined in (g) and (h)?

$$P = .154646 \qquad \text{they are about the same}$$

Discussion: Probabilities can be computed either in terms of the original normal distribution model or in terms of the standard normal curve. Using the z-scores provides the additional information of how many standard deviations an observation lies from the mean.

(o) A height of 151.8 cm is how many standard deviations below the average female height? What is the probability of randomly selecting a woman who is at most 152 cm tall?

z-score: 2

probability below: .02275

(p) How short would a man have to be to be in the same percentile as a 152 cm tall woman? [*Hint*: In the Minitab menu, change to the "inverse cumulative probability" option and enter the probability as the input constant. Minitab will report the corresponding x value.]

Discussion: Notice in this last problem that you were given the probability and asked to find the corresponding height. This "reverses" the earlier calculations where you were given the height(s) of interest and asked to find the corresponding probability.

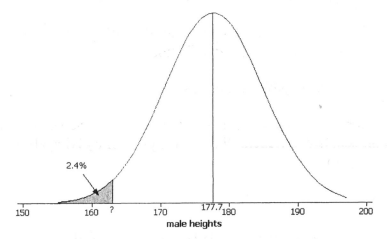

2.4%

150   160   ?   170   177.7 180   190   200

male heights

(q) What is the *z*-score for the observation in (p) and how does it compare to the *z*-score in (o)?

Discussion: You should have found that both observations are roughly 2 standard deviations below the mean, putting them in about the 2.5th percentile. This agrees with what the empirical rule would have predicted since approximately 2.5% of observations will fall more than 2 standard deviations below the mean.

Investigation 4-5: Birth Weights (Applet Exploration)

Birth weights of babies in the United States have been said to be reasonably modeled by a normal distribution with mean 3250 grams and standard deviation 550 grams, N(3250, 550). Babies weighing less than 2500 grams are considered to be of low birth weight.

(a) Produce of a sketch of the normal density curve described by these parameters (with appropriate scaling). Provide a label and a scale for the horizontal axis.

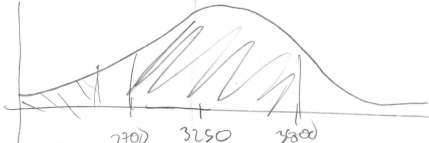

2700   3250   3800

(b) Approximately 68% of newborns weigh between which two values? Shade the corresponding area of the distribution above.

2700 , 3800

(c) On the graph in (a), shade the area of the distribution corresponding to babies of low birth weight. Estimate from your shading the percentage of the distribution that falls in this range.

13%

To estimate the probability of a randomly selected baby being of low birth weight, we will use the normal model to determine P(X≤2500) where X, which denotes the birthweight of a randomly selected baby, follows a N(3250, 550) distribution. We can calculate these probabilities either using Minitab or an applet. The applet is slightly more visual.

(d) Open the "Normal Probability Calculations" applet and specify 3250 as the mean and 550 as the standard deviation. Press the "Scale to Fit" button. Select the first row and in the X box enter the value of 2500 and hit the Enter key. [Note the applet scales both the *x*-axis and the "*z*-axis."]

What does the applet report for the *z*-score and the probability?

-1.36

Write a one-sentence interpretation of this *z*-score and this probability.

Normal Probability Calculations

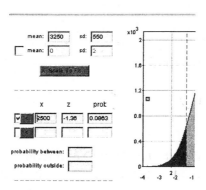

They are a little more than
1 σ away from m.

(e) Data from the *National Vital Statistics Report* indicate that there were 3,880,894 births in the United States in 1997. A total of 291,154 babies were of low birth weight. What is the observed proportion of low birth weight babies and how does this compare to the proportion predicted by the normal model?

.075   slightly less than expected

(f) Suppose you wanted to estimate the probability of a baby weighing more than 10 pounds (4536 grams) at birth. Sketch (and label) the distribution and shade the area of interest.

(g) In the applet, change the inequality to >, specify 4536 in the *X* box, and press Enter. What are the *z*-score and the probability reported by the applet? Confirm that these numbers appear consistent with your sketch.

2.34   &   .0097

(h) Suppose you wanted to estimate the probability that a randomly selected baby weights between 3000 and 4000 grams. Sketch (and label) the probability model with the area of interest shaded.

3000   4000

(i) In the applet, specify 3000 in the first *X* box and then select the second row and specify 4000 in the second *X* box. Press the Enter key. What does the applet report for the *z*-scores and probability between these two values?

-.45 | 1.36

prob between : .5873

(j) Data from the *National Vital Statistics Report* indicate that there were 2,552,852 newborns weighing between 3000 and 4000 grams in 1997 How closely does the model's prediction in (i) match the observed relative frequency?

.6578   which is slightly more than expected

(k) You can also use the normal distribution model to "work backwards." Suppose you want to know how much a baby would have to weigh to be among the lightest 2.5% of all newborns (the 2.5th percentile). Sketch (and label) a normal model and indicate this area on the sketch.

2.5th

(l) In the applet, unselect the second row. Since we want the *lightest* 2.5%, make sure the first row inequality is set at < and change the probability box value to .025. Hit the Enter key. The applet will determine the corresponding *z*-score and weight. Are the results consistent with your sketch?

2172.3 $z = -1.96$

(m) Use the applet to determine how much a baby would have to weigh at birth to be in the heaviest 2.5% of all newborns. How does this *z*-score compare to that in (l) and how do the weight cut-offs compare? Explain these relationships.

4327.9   1.96

they, are the same distance from μ

(n) Suppose instead that birth weights of newborns followed a normal distribution with mean 3500 and standard deviation 400. Use the applet to create this sketch (pressing the "Scale to Fit" button) and to determine the two values such that the middle 95% of birth weights fall in between the two values.

2700 , 4300   2716
2 & 2   4284

1.96   1.96

(o) How do the *z*-scores in (n) compare to those in (l) and (m)?

z-scores are approx the same

same

In working with a normal distribution model, N(μ, σ), approximately 95% of observations will fall within 2 standard deviations of the mean (i.e., within $\mu \pm 2\sigma$), regardless of the values of μ and σ. This is the theoretical basis of the *empirical rule* discussed in Ch. 2.

Discussion: You should have found that you can determine probabilities for a wide variety of intervals, e.g., P(X$\leq$x), P(X$\geq$x) and P(a$\leq$X$\leq$b). You can calculate these probabilities either with Minitab or with this applet. When using Minitab, you will start with P(X$\leq$x) and then perhaps will need to do a little arithmetic to find the probability of interest. You can also invert the process and find the value of x that achieves a certain probability. Keep in mind that these probabilities are only as accurate as the model.

Practice:

4-8) Birth Weights (cont)
For the N(3250, 550) model of newborn weights:
(a) How much would a baby have to weigh to be among the heaviest 10% of all newborns?
(b) Between what two values do the middle 80% of all birthweights fall?
(c) Could the empirical rule have been used to answer (b)? Explain.

4-9) Heights (cont.)
According to the National Health Survey, heights of American males are considered to be normally distributed with mean μ=69 in and standard deviation σ=2.8in.
 (a) The US Marine Corps require that men have heights between 64 in and 78 in. Find the percentage of men meeting that height requirement. [Make sure you include a well-labeled sketch of the distribution and the areas of interest. Include any Minitab output of applet screen capture.]
(b) The Beanstalk Club is a social organization for tall people. Men are eligible for membership if they are at least 74 in tall. What is the probability that a randomly selected American male is eligible for the Beanstalk Club?
(c) If the top 5% of women are eligible and women's heights are normally distribution with mean μ = 63.6 in and standard deviation σ = 2.5 in, how tall does a woman need to be to be eligible?

4-10) Paternity Debates

An expert witness in a paternity suit testifies that the length (in days) of pregnancy (the time from conception to delivery of the child) is approximately normally distributed with mean $\mu = 270$ days and standard deviation $\sigma = 10$ days. The defendant in the suit is able to prove that he was out of the country during a period that began 280 days before the birth of the child and ended 230 days before the birth.

(a) If the defendant was the father of the child, what is the probability that the mother could have had the very long or the very short pregnancy indicated by the testimony? [Make sure you include a well-labeled sketch of the distribution and the areas of interest. Include any Minitab output or applet screen capture.]

(b) If you were the judge in this case, based on the calculation in (a) alone, how would you rule? Explain.

SECTION 4-3: DISTRIBUTIONS OF SAMPLE COUNTS AND PROPORTIONS

The normal distribution provides us with a very powerful model for how many real-world variables are distributed. But a second, perhaps even more important, reason for its widespread use in statistics is that it is also a powerful tool for modeling the long-term behavior of how many sample statistics vary from sample to sample. In other words, the normal model effectively approximates the *sampling distribution* of many sample statistics in many circumstances. This can greatly aid our ability to make predictions for what we expect to see in our sample results, and also to make inferences about the population based on sample results—as long as the model applies.

Investigation 4-6: Reese's Pieces (Applet exploration)

Each student in your class will take a random sample of $n=25$ Reese's Pieces candies from the population of all candies manufactured by Hershey.
(a) Let X = number of orange candies in the sample. Does X satisfy the conditions to be (approximately) a binomial random variable? Explain.

(b) Do you think every sample will have the same number of orange candies? Explain.

(c) Record the number of orange candies in your sample. Is this number a parameter or a statistic? Explain.

(d) Combine your results with the rest of the class and produce a dotplot of the distribution of the number of orange candies in these samples. Remember to label and scale the horizontal axis.

(e) Did everyone in the class obtain the same number of orange candies?

(f) Describe the pattern of variation in these results, including an approximation of the average number of orange candies observed.

(g) Does it appear that a normal density curve could reasonably model the behavior of this distribution? Explain. Approximate the mean and standard deviation of this distribution.

(h) How would the above dotplot change if we had recorded the *proportion* of orange candies x/25 in each sample instead of the count? Also approximate the mean of this distribution.

To better understand this pattern of variation, we need to take many more samples of 25 candies each. We will turn to technology to simulate this process. To perform the simulation, we need to tell the computer the value of the population proportion of orange candies. Suppose that 45% of the candies produced by this manufacturer are orange. We will denote this population proportion by $\pi = .45$.

Reese's Pieces Samples

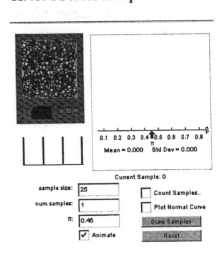

(i) Open the "Reese's Pieces" applet. Note that we have set the applet to assume that 45% of the population of Reese's Pieces candies is orange.
The applet is also set to take a sample of $n=25$ candies to match your class activity.
Click the Draw Samples button. The applet will randomly select a sample of 25 candies and report the sample proportion of orange candies, denoted by $\hat{p}$.

Click the Draw Samples button again.

Did you obtain the same value for $\hat{p}$ each time?

What is the average of the two $\hat{p}$ values you obtained?

What is the standard deviation of the two $\hat{p}$ values obtained? [This is reported by the applet underneath the dotplot.]

(j) Change the number of samples from 1 to 500, uncheck the Animate box, and click the Draw Samples button again. Describe the shape, center, and spread of the resulting empirical sampling distribution of $\hat{p}$ values.

(k) Check the Plot Normal Curve box to overlay a normal density curve. Does the normal distribution appear to be an appropriate model for this distribution of sample proportions?

The distribution of sample proportions appears to be reasonably modeled by a normal distribution with mean .45 and standard deviation close to .1.

(l) Assuming that this normal approximation is valid, what percentage of the sample proportions does the empirical rule predict will fall in the interval (.35, .55)? In the interval (.25, .65)? In the interval (.15, .75)?

    (.35, .55):   (.25, .65):   (.15, .75):

(m) Click the "Count samples…" box, change the menu from "count above" to "count between" and enter .35 in the above box and .55 in the below box. Click OK and the applet reports the percentage of the 500 sample proportions that are in the interval (.35, .55). Similarly, determine the percentage that falls in the interval (.25, .65) and the interval (.15, .75). [*Hint*: Use the mouse to drag the red vertical lines.] What are these three percentages?

    within $\pm$.10 of .45:   within $\pm$.20 of .45:   within $\pm$.30 of .45:

(n) How do these match up with the percentages predicted by the empirical rule?

(o) If we had instead taken samples of size $n=75$, how do you think the sampling distribution of the sample proportions would compare to the distribution when $n=25$? Explain.

(p) In the applet, change the sample size to $n=75$ candies in each sample and draw 500 samples.
- Does the normal probability model still appear to be appropriate?
- What is the main difference in comparing this new sampling distribution to the previous one?
- What percentage of the sample proportions fall within $\pm$ .10 of π, .45?

(q) If the population was instead 65% orange, how do you think the sampling distribution of the sample proportions would change?

(r) Change π from .45 to .65, keeping the sample size at $n=75$, and use the applet to obtain an empirical sampling distribution. Describe this new distribution and discuss whether your prediction in (q) was correct.

(s) Suppose we had taken samples of size $n=5$ instead. Predict how the sampling distribution will change and then use the applet to check your prediction. Discuss your observations.

Discussion: In this investigation you first considered the random variable X = number of orange candies in a sample of $n=25$ candies. If we assume that the candies are coming from an infinite population/process, then X follows a binomial distribution with parameters $n=25$ and π equal to the proportion of orange candies produced by this manufacturer. In the above investigation you found that when n is sufficiently large, the sampling distribution of the sample proportion of successes $\hat{p} = X/n$ is well modeled by a normal distribution. We will consider this approximation to be valid whenever $n\pi \geq 10$ and $n(1-\pi) \geq 10$, but this is just a rough guideline and the approximation becomes more accurate as the sample size increases. To apply this normal model, we need to know the mean and standard deviation of $\hat{p}$, so in the next probability detour you will determine the theoretical mean and standard deviation analytically.

An analogous result is that when n is sufficiently large, the distribution of the sample count X, which follows a binomial distribution, is also well approximated by a normal distribution. We will refer to this as the *normal approximation to the binomial*. One thing to keep in mind is that a normal distribution is *continuous* (in principle it can take on all possible values in any interval of real numbers) but the distributions of sample counts (or proportions) are *discrete* since the values must be integers (or multiples of $1/n$). Still, it appears a normal probability model could reasonably describe this distribution for large values of n (when this discrete nature is less apparent). We can use a continuity correction (see Practice 4-13) to make the normal approximation even more accurate with moderately small values of n.

If the population is not infinite, then we need the population to be much larger than the sample so that the binomial distribution reasonably approximates the hypergeometric distribution. Recall that our rough guideline is to consider this approximation valid whenever the population size is at least 20 times larger than the sample size.

Probability Theory Detour: Mean and Standard Deviation of $\hat{p}$

In chapter 3, your were told that the expected value of a binomial random variable with parameters n and π was equal to $n\pi$ as your intuition should confirm. You were also reminded that this "expected value" represents the long-run average over many, many observations of the random variable.

(a) If $n = 25$ and $\pi = .45$, calculate the expected number of candies in a random sample. [Recall that this expected value does not need to be an actually obtainable value.]

(b) If 11.25 is the expected value of X, then what do you conjecture for the expected value of $\hat{p}$?

Probability rule: If c is some constant and we multiply our random variable by c, we determine the expected value of this new quantity through the relation: $E(cX) = cE(X)$

(c) Use this probability rule to determine an expression for $E(\hat{p})$, that is, $E(X/n)$, in terms of n and/or π. [Hint: Use the probability rule to get $E(\hat{p})$ in terms of $E(X)$ which you know.]

Similarly, we can derive an analytical expression for the standard deviation of $\hat{p}$. Recall that the standard deviation of a binomial random variable is $\sqrt{n\pi(1-\pi)}$.

Probability rule: If c is some constant greater than zero and we multiply our random variable by c, we can determine the standard deviation of this new quantity through the relation:
$$SD(cX) = cSD(X).$$

(d) Use this probability rule to determine an expression for $SD(\hat{p})$ in terms of n and/or π.

Thus, if we have a sample proportion determined from a binomial random variable (independent observations, all with π as the probability of success) with parameters n and π, then the $E(\hat{p}) = \pi$ and $SD(\hat{p}) = \sqrt{\pi(1-\pi)/n}$.

(e) If $n = 25$ and $\pi = .45$, calculate the expected value and standard deviation of the random variable $\hat{p}$ from this process. Are these results close to the empirical average and expected value reported by the applet?

$E(\hat{p})$:   Applet:

$SD(\hat{p})$:   Applet:

The expected value and standard deviation reports are always valid (assuming we have a random sample). Futhermore, the shape of the sampling distribution of $\hat{p}$ will be approximately normal when the normal approximation to binomial is valid.

In summary:

The shape of the sampling distribution of the sample proportions will be approximately normal *if* $n\pi \geq 10$ and $n(1-\pi) \geq 10$.

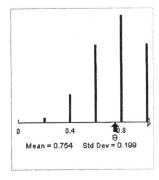

When n was 5 and π was .75, these conditions were not met, and the distribution was skewed and very granular.

However, when n was 25 and π was .45, these conditions were met and the normal model approximated the distribution closely.
This result holds whether we are sampling from a process or a large population.

Central Limit Theorem (CLT) for a Sample Proportion:
Suppose that a random sample of size n is taken from a process or a large population (whose size is at least $20 \times n$) in which the probability of successes is π. Then the sampling distribution of the sample proportion $\hat{p}$ can be approximately modeled by a normal probability curve with mean equal to π and standard deviation equal to $\sqrt{\pi(1-\pi)/n}$. This shape becomes more and more normal as the sample size n increases, and it is generally considered to be valid provided that $n\pi \geq 10$ and $n(1-\pi) \geq 10$.

Discussion: While in many situations we can apply the hypergeometric or the binomial distributions to obtain exact probabilities, the above theorem allows us to approximate these probabilities using a normal distribution. Before computers became so widespread, it was much easier to perform normal probability calculations, for which tables of values were readily available, than binomial or hypergeometric probability calculations. We will still often make use of this relationship and calculate an approximate normal probaiblity instead of the exact binomial or hypergeometric probability when the "technical conditions" of the Central Limit Theorem (n $\pi \geq 10$ and $n(1-\pi) \geq 10$) are satisfied. In either case, we must always start with a random sample.

(f) Use the N(.45, $\sqrt{.45(1-.45)/25}$) model to approximate the probability of a random sample of 25 candies having more than 50% orange candies. *Hint*: Always start with a well-labeled sketch of the distribution.

(g) Use the appropriate model to approximate the probability of a random sample of 75 candies having more than 75% orange candies. How does this probability compare to the probability in (f)? Is this what you expected? Explain.

(h) Use the normal model to approximate the probability that the sample proportion of orange candies in a random sample of 75 candies will fall between .35 and .55.

Practice:

4-11) ESP (cont.)

Suppose you are giving someone an ESP test with 4 possible symbols to match. You plan to give them a test of 40 questions. Let X represent the number of correct answers.
(a) Does the Central Limit Theorem apply here?
(b) According to the Central Limit Theorem, sketch and describe the shape, center, and spread (specify the mean and standard deviation) of the sampling distribution of sample proportions.
(c) Use the model in (b) to determine the z-score for an observed proportion of .375.
(d) Use the model in (b) to approximate the probability that a subject gets at least 37.5% correct responses (using either Minitab or the Normal Probability Calculator applet).

4-12) Heart Transplant Mortality (cont.)

Recall Investigation 3-10 in which you examined whether the heart transplant mortality rate at St. George's Hospital in London exceeded the 15% benchmark rate. Suppose 15% of heart transplant patients die. You plan to take a random sample of 361 patients and calculate the proportion of these patients who die.
(a) Does the Central Limit Theorem apply here?
(b) According to the Central Limit Theorem, sketch and describe the shape, center, and spread (specify the mean and standard deviation) of the sampling distribution of sample proportions.
(c) Using the model in (b), calculate the z-score for an observation of $\hat{p} = .197$.
(d) Use the model in (b) to approximate the probability of a sample proportion of deaths of .197 or higher in a random sample of 361 patients from this process (using either Minitab or the Normal Probability Calculator applet).

4-13) Continuity Correction

Reconsider the Reese's Pieces situation where $n = 25$ and $\pi = .45$. Let X represent the number of orange candies in a randomly selected sample from this process. Recall that the mean of a binomial random variable is $n\pi$ and the standard deviation is $\sqrt{n\pi(1-\pi)}$.
(a) Use the binomial distribution to determine the exact probability of obtaining at most 10 orange candies in a random sample this size.
(b) Use the normal approximation to the binomial distribution to approximate the probability of at most 10 orange candies in a sample. How does your answer to compare to (a)?
(c) Using the Normal model in (b), how does P(X≤10) compare to P(X<10)?
(d) Since the Normal probability curve is continuous, we can use a "continuity correction" to improve this approximation. Instead of finding the probability of at most 10 candies, we can find the probability of at most 10.5 candies (half-way between 10 and the next possible value of 11) in a normal distribution to ensure we include the "mass" near X=10. Repeat (b) using 10.5 as the cut-off value of interest.
(e) Use the Normal model indicated by the Central Limit Theorem for a sample proportion, *with an appropriate continuity correction*, to approximate the probability of obtaining a sample proportion of at most .40. Make sure you state what probability you calculated.

Investigation 4-7: *Cohen v. Brown University*

In 1991, a suit was filed against Brown University after Brown terminated funding for its women's gymnastics and volleyball teams and its men's water polo and golf teams. The suit charged that Brown was violating Title IX of the Education Amendments of 1972, the federal law that prohibits sex discrimination by all educational institutions receiving federal funds. This requires men and women to have equivalent opportunities for participation. A main component of the plaintiff's case was that while 51% of the undergraduate student body was women, only 38% of the 897 students engaged in intercollegiate athletics were women.

If there is no gender discrepancy, then Title IX assumes that the proportion of women athletes should be similar to the proportion of women in the overall student body. However, we know the sample result can deviate from this population proportion "just by chance." Suppose we were randomly selecting students to be athletes, the question is whether this random process could lead to such a disparity in these proportions. While we know the proportion of women in the population of all university students to be .51, we don't know the probability of an athlete being female. Let π refer to the probability of a Brown university athlete being female.

(a) What are the observational units and variable of interest in this study? Also specify the population or process of interest. [Note: While we have not actually taken a random sample here, we will assume the sample of 897 students is representative of this overall process.]

Observational units: *props of women doing parts   female?*

Population/process of interest: *athletes*

Unknown parameter of interest: *prob of Brown athlete = female*

(b) State a null and an alternative hypothesis about the value of π (in symbols and in words).

H$_0$:

H$_a$:

(c) We can assess the strength of evidence against the null hypothesis by treating the 897 current athletes as a random sample from the process of athlete determination at Brown. If we initially give the university the benefit of the doubt and assume the null hypothesis is true, does the Central Limit Theorem for a sample proportion apply here? Explain. (*Hint*: Use the hypothesized value of π to check the conditions.)

(d) If $\pi = .51$ as hypothesized, how many standard deviations is the observed sample proportion ($\hat{p} = .38$) from the hypothesized probability of success (.51)? [*Hint*: Use the hypothesized value of π in calculating the standard deviation of the sample proportion.]

(e) If $\pi = .51$, use the Central Limit Theorem to approximate the probability that at most 38% of the sample would be female. (*Hint*: Include a well-labeled, scaled, and shaded sketch of the relevant normal probability model and the area of interest. Then use Minitab to calculate the normal probability. You need not use the continuity correction with such a large sample size.)

(f) Based on this *approximate p-value*, what conclusion do you draw about whether this discrepancy could have arisen by chance? In other words, does your analysis suggest that the proportion of women involved in intercollegiate athletics is significantly lower than the proportion of women students at the university? Explain.

Study Conclusion: The sample proportion indicates a much lower percentage of women among athletes. The p-value again tells us how often "our data" would occur "by chance." In this chapter, the chance mechanism is the random sampling process, assuming the null hypothesis is true ($\pi = .51$). The p-value for this test is extremely small, providing very strong evidence that this discrepancy did not occur by random sampling variation alone (assuming this sample is representative of the process at Brown University). Therefore, we have strong evidence that this sample did not come from a process with $\pi = .51$. This indicates that the proportion of women among athletes is less than we would expect based on the proportion of women students at the university. This analysis is valid as long as the data from this year are representative of the overall process at Brown University. We are not conducting an experiment, so we are not making any statements as to why π is below .51. In fact, this level of the Title IX legislation is not taking into considering the interest level of the women at the university.

TEST STATISTIC
Minitab will only report this approximate p-value to four decimal places. Clearly it is very small, and the *z*-score computed in (d) tells us that the observed sample proportion is more than 7 standard deviations from the hypothesized value!

The calculation you carried out in (d) is referred to as the *test statistic*, which provides a standardized measure of the distance between our observed result and our hypothesized value. The p-value indicates how often we would obtain a test statistic at least this extreme when the null hypothesis is true.

Since we are working with a normal distribution, we know that an observation more than 2 or 3 standard deviations away from the mean is a surprising outcome.

In general, to test $H_0: \pi = \pi_0$, where π_0 indicates the hypothesized value of the population proportion, the test statistic is obtained through the formula:

$$z_0 = \frac{observed - hypothesized}{standard\ deviation} = \frac{\hat{p} - \pi_o}{\sqrt{\pi_o(1-\pi_o)/n}}$$

and the p-value calculation will depend on the form of H_a:

If $H_a: \pi > \pi_0$, the p-value is $P(Z \geq z_0)$.

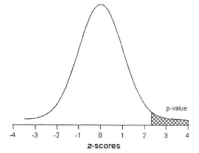

If $H_a: \pi \neq \pi_0$, the p-value is $2P(Z \geq |z_0|)$ – a "two-sided p-value"

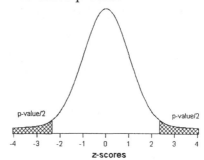

If $H_a: \pi < \pi_0$, the p-value is $P(Z \leq z_0)$.

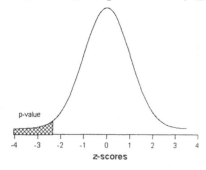

Question: Why are we able to assume the p-value is equally divided on each side here?

We ultimately find this p-value using either a *normal probability table* or technology (either a computer or calculator). If you are using the normal distribution directly, you can find the probability either in terms of $\hat{p}$'s distribution or in terms of the test statistic and the standard normal distribution.

Discussion: You have now seen three ways to calculate a p-value in the binomial setting.

1) Approximate the p-value through simulation. This method has the advantage of emphasizing what a p-value represents: the fraction of times that such a result at least this far from the hypothesized value would occur by chance alone if the process were repeated over and over. Another advantage is that simulation can be used as a very flexible tool for approximating a p-value in almost any setting. One obvious disadvantage is that the p-value produced is only approximate. Another potential disadvantage is that simulation requires computer software to produce enough repetitions for the approximation to be reasonably accurate.

2) Calculate the exact p-value using the binomial probability model. This method has the virtue of producing an exact p-value. A disadvantage is that calculating the p-value can be cumbersome without technology, particularly when n is large. Another potential disadvantage, as compared to simulation, is that this method does not directly relate to the long-run interpretation of a p-value.

3) Approximate the p-value using the normal approximation to the binomial distribution. An advantage of this method is that calculating a z-score (test statistic) is not cumbersome, and this test statistic provides a nice measure of how far (how many standard deviations) the observed value is from the hypothesized value. This is a flexible idea that applies in many settings. However, to then convert the test statistic to a p-value requires a normal probability table or technology. Other disadvantages are that this method produces only an approximate p-value, and the approximation produced is only reasonable when certain conditions are satisfied (the validity of the normal approximation to the binomial).

TESTS OF SIGNIFICANCE STRUCTURE

The above investigation outlined steps that you will follow in general in carrying out a test of significance.

1) ***Define the population parameter in words*** - what is the unknown quantity that you are trying to make a decision about?

2) ***State the null and alternative hypotheses about this parameter***—we will always write the null hypothesis as "parameter = value" and the alternative hypothesis will be the same parameter symbol, the same value, but then you will choose between <, >, and ≠depending on the research question. This choice should be made before you look at the data.

3) ***Check the "technical conditions" and sketch the sampling distribution***—in the last chapter you verified whether the sampling process was binomial and to use the methods in this chapter you also have to first verify whether the normal approximation to the binomial is valid (the conditions of the Central Limit Theorem).

4) ***Calculate the test statistic (normal-based inference)***—this provides a measure of the distance between the sample result and the hypothesized value of the parameter, often in terms of "how many standard deviations" apart.

5) ***Calculate the p-value***—this can be done directly from the sample result or from the test statistic and provides a measure of how likely you are to find a sample result at least this extreme (at least this many standard deviations away) when the null hypothesis is true, by chance alone.

6) ***Make a decision about the null hypothesis***—you will either "reject H_0" or "fail to reject H_0" depending on the size of the p-value and the level of significance.

7) ***State your conclusion in context***—you should never stop at step 6, but need to go back and answer the researcher's question (e.g., there is significant evidence that the proportion of women among athletes is smaller than the proportion of women among the general population at this university).

Investigation 4-8: Kissing the Right Way (cont.)

Recall the study that analyzed a sample of 124 couples and found 80 turned to the right when kissing. Suppose we take a random sample of 124 kissing couples and let X represent the number of couples in the sample who turn to the right. You saw in chapter 3, that X can be considered a binomial random variable with parameters $n=124$ and π =probability of a couple turning to the right. Earlier you tested the Dr. Güntürkün's conjecture through the hypotheses:

H_0: $\pi = 2/3$ (the probability of turning to the right is 2/3)

H_a; $\pi \neq 2/3$ (the probability of turning to the right differs from 2/3)

using the binomial distribution (two-sided binomial p-value = .634, 95% CI for π=(.56,.73)).

(a) Does the Central Limit Theorem apply here? Explain.

(b) If so, what does the Central Limit Theorem say about the sampling distribution of sample proportions? (Comment on shape, center, and spread, and include a sketch of this distribution. Keep in mind that we are assuming the null hypothesis to be true when creating this sampling distribution.)

(c) Using the sample data obtained by Dr. Güntürkün, shade the area of interest in the sampling distribution of the sample proportion, calculate the test statistic and the *two-sided* p-value for this test, decide whether you should reject or fail to reject the null hypothesis at the .05 significance level, and state your final conclusion about whether the probability a couple turns to the right when kissing differs from 2/3 as conjectured.

(d) How does the p-value in (c) compare to that found earlier with the binomial distribution?

(e) Provide an interpretation for the test statistic in the context of this problem.

(f) How small would the sample proportion of $\hat{p}$ need to be before we would reject this null hypothesis at the 5% level? [*Hint*: Use the inverse cumulative probability to answer this reverse question. You might also want to think of this problem in terms of the necessary test statistic value. Like always, start with a sketch.]

(g) How large would the sample proportion need to be before we would reject this null hypothesis at the 5% level?

(h) Repeat (f) and (g) for the 1% level of significance. How do these cut-offs compare to those found in (f) and (g)? Explain why this makes sense.

TYPE I AND TYPE II ERRORS

Def: The *rejection region* of a test of significance is the values of the sample statistic that lead us to reject the null hypothesis. The definition of the rejection region will depend on the level of significance specified.

In the above example, for a 5% level of significance, the rejection region would be the values of $\hat{p}$ that are below .584 or above .750.

These are the values that are far enough from 2/3 that we would obtain a two-sided p-value $\leq .05$ and would therefore reject the null hypothesis (that π = 2/3) at the 5% level (see figure). If the level of significance had been 1%, then the rejection region would have been: $\hat{p} \leq .558$ and $\hat{p} \geq .776$. This rejection region is smaller, as we are requiring the sample result to be even more extreme in order to convince us to reject the null hypothesis.

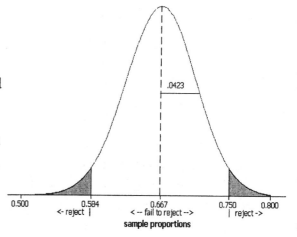

(i) For a 5% level of significance, what is the probability that an observation will fall in the rejection region when the null hypothesis is true? Is this a Type I or a Type II Error?

(j) Suppose π was actually equal to .50 (that kissing couples turn to the right 50% of the time). What is the probability that we will (correctly) reject the null hypothesis that $\pi = 2/3$ at the 5% level? [*Hint*: What is the behavior of the sampling distribution with $\pi = .50$? How often do we obtain a $\hat{p} \leq .584$ or $\geq .750$ with this sampling distribution?]

(k) Repeat (i) and (j) for the 1% level of significance.

P(reject H_0) when $\pi = 2/3$:

P(reject H_0) when $\pi = 1/2$:

Discussion:

In (j), you found the *power* of the test against the alternative value of .5 by finding P($\hat{p} \leq .5844$ or $\hat{p} \geq .7499$ assuming $\pi = .5$) = .9693. This means that if π actually equals .5 and we repeatedly sample from this population (or process), we will reject H_0: $\pi = 2/3$ in about 97% of samples.

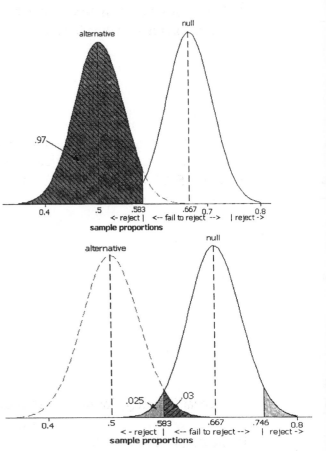

As you saw in chapter 3, the probability of a Type I Error is equal to the level of significance. The probability of a Type II Error is 1 minus the power, so in this case P(Type II Error) = 1-.9693=.0307. We could have found this directly by finding P(.584 $\leq \hat{p} \leq$ .750 when $\pi = .5$).

In order to do these calculations, you must first determine the *rejection region* based on the null hypothesis and the level of significance, and then see how often you are in (or not in) this rejection region for a particular alternative value of the parameter.

(l) Discuss whether the power of the test will be larger, the same, or smaller for the following scenarios [*Hint*: Sketches should help! Note that the change in each scenario appears in bold.]:

1. $n = 124$, $\alpha = \mathbf{.10}$, $\pi_a = .5$

2. $n = \mathbf{250}$, $\alpha = .05$, $\pi_a = .5$

3. $n = 124$, $\alpha = .05$, $\pi_a = \mathbf{.6}$.

SAMPLE SIZE DETERMINATION

Typically in designing studies, statisticians will set the level of significance (controlling the probability of a type I error) and then determine the sample size necessary to achieve a certain level of power against an alternative value of interest (controlling the Type II error rate).

(k) Suppose researchers want to test $\pi = 2/3$ and want the P(Type II Error when $\pi = .5$) to be at most .01. Determine the sample size necessary. [*Hints*: How does changing n affect the bounds of the rejection region? If this probability is .01, how many standard deviations apart are the lower bound of the *rejection region* and the actual value of π, .5? Express this as a function of n and solve for n.]

Study Conclusions: The sample proportion turning to the right was .645, below the 2/3 probability conjectured by Dr. Güntürkün. However, the sample data do not provide convincing evidence against Dr. Güntürkün's conjecture a kissing couple turns to the right 2/3 of the time (two-sided p-value = .645 indicating that we would see a sample proportion at least this extreme in 64.5% of random samples when $\pi = 2/3$). Since the power of this test against the alternative of $\pi = .5$ is reasonably large (if the sample was coming from a process with $\pi = .5$, we would reject the null hypothesis in almost 97% of random samples), this gives us even more reason to think that we had a large enough sample size to detect a difference of this size had there been one. While measures were taken to ensure a representative sample of different environments (airports, train stations, beaches, and parks) and cultures (US, Germany, and Turkey), this was not a true random sample and we should be cautious in generalizing these conclusions.

Discussion: The normal approximation, when valid, provides the same information as the binomial calculation in chapter 3. The conclusions above are essentially the same as those made in earlier with the binomial distribution. There are some advantages to using the normal approximation, such as providing the test statistic as an additional measure of the discrepancy, and in being able to answer carry out power calculations and sample size determinations. You will see that the calculations for a confidence interval are also greatly simplified.

Investigation 4-9: *Cohen v. Brown University* **(cont.)**

Reconsider Investigation 4-6 in which you analyzed whether sample data provided strong evidence that π, the proportion of females among Brown University athletes was less than .51, the proportion of females among all undergraduates at the university. You found a very small p-value and so overwhelmingly rejected .51 as a plausible value for π. A natural follow-up question is often "then what are the plausible values of π?"

In the last chapter, we defined a confidence interval as the collection of plausible values of a parameter, where "plausible" means that the observed sample result would not be surprising given that parameter value. You found this interval by using mostly trial and error to see which values of π lead to a two-sided p-value of at least .05. Due to the discrete nature of the binomial distribution, you did not find values that lead to a p-value of exactly .05. Using the normal probability model, you can *solve* for an appropriate cut-off value.

(a) Suppose that we take repeated random samples of size n from a population (or process) where the proportion (or probability of) having the characteristic of interest is π. According to the empirical rule, 95% of the sample proportions should be within what distance of π?

(b) Thus, in 95% of samples, π should be within what distance of the sample proportion?

(c) If π is unknown, suggest a way of approximating the standard deviation of the sample proportions.

Def: An approximation of the standard deviation of a sample statistic is often referred to as a *standard error*. A standard error is calculated from the sample data itself. We denote the standard error of a sample proportion by SE($\hat{p}$).

(d) Use your suggestion in (c) to calculate standard error for the sample proportion of females in this study [recall that the study sample size was 897 and that 38% of the sample was female].

(e) What values fall within 2 standard errors of $\hat{p}$ = .38?

You have approximated a 95% confidence interval for π, those values that fall within about 2 standard deviations of $\hat{p}$. An approximate 99.7% confidence interval would specify the values that fall within about 3 standard deviations of $\hat{p}$ as the plausible values for π. These intervals are approximate because we are approximating the standard deviation, because we are approximating the binomial distribution with the normal probability model, and because the empirical rule provides rough approximations as well.

To address the empirical rule approximation, let's find the value $z*$ so that $P(-z* \le Z \le z*)=.95$ more exactly.

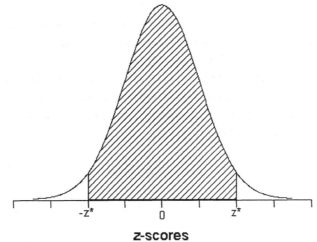

z-scores

(f) What percentile of the standard normal distribution does $z*$ correspond to? In other words, what is the total area to the left of the value $z*$ in the above sketch?

(g) Use Minitab's invcdf command (or, equivalently, use Calc> Probability Distributions> Normal with the "inverse cumulative probability" button) to determine $z*$ to 2 decimal places.

(h) What values are within this many standard errors of the sample proportion $\hat{p}$ =.38?

(i) Is .51 inside the 95% confidence interval for pi? Explain why this makes sense based on your earlier analysis.

An approximate **C% confidence interval for a population proportion or process probability**
π can be calculated using the expression:
$$\hat{p} \pm z^* \sqrt{\hat{p}(1-\hat{p})/n}$$
where $-z^*$ is the (100-C)/2 percentile of a *standard normal* probability model. This procedure is
considered valid when the data are randomly sampled from the population/process of interest and
$n\hat{p} \geq 10$ and $n(1-\hat{p}) \geq 10$, where $-z^*$ is the (100-C)/2 percentile of a *standard normal*. This
procedure is sometimes referred to as the *Wald interval*. It provides an alternative to the exact
binomial confidence interval that you studied in Ch. 3.

The *half-width* of the interval $z^* \sqrt{\hat{p}(1-\hat{p})/n}$ is also referred to as the *margin of error*.
We can write the interval as *estimate* $\pm$ *margin of error*, or as *estimate* $\pm$ *critical
value*×*SE(estimate)*, both of which will be common forms for confidence intervals that estimate
other parameters, as you will study later.

We will interpret this interval as the plausible value for the parameter based on the
observed sample statistic. There is a *duality* between the confidence interval and *two-sided* tests
of significance: If a hypothesized value is rejected at the α level of significance, then it will not
be included in the $100(1-\alpha)\%$ confidence interval.

(i) Write a one sentence interpretation of the confidence interval from (h) in the context of this
study.

(j) Repeat this analysis to determine the value of z^* for a 99% confidence interval to 2 decimal
places and calculate the 99% confidence interval for π in the kissing study. How does this
interval compare to the one in (h)?

(k) Verify the above calculations with Minitab by choosing Stat > Basic Statistics > 1 Proportion
as before but under the Options button, select the box next to "Use test and interval based on
normal distribution."

Study Conclusions: Our earlier analysis gave us strong evidence that π was smaller than .51.
We can now also say that we are 95% confident that the probability of an athlete being female is
between .348 and .412. Further, we are 99% confident that the underlying probability for this
process is between .338 and .422.

Summary of One-Sample z Procedures for Proportion

Test of H_0: $\pi = \pi_0$
Technical conditions: Random sample and $n\pi_0 \geq 10$ and $n(1-\pi_0) \geq 10$ (which means we expect to see at least 10 success and 10 failures in the sample)
Test statistic: $z_0 = (\hat{p} - \pi_0)/\sqrt{\pi_0(1-\pi_0)/n}$
p-value:

If H_a: $\pi > \pi_0$, the p-value is $P(Z \geq z_0)$.
If H_a: $\pi < \pi_0$, the p-value is $P(Z \leq z_0)$.
If H_a: $\pi \neq \pi_0$, the p-value is $2P(Z \geq |z_0|)$

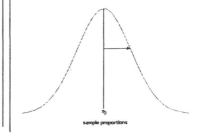

C% confidence interval for π (Wald):
Technical conditions: Random sample/process and $n\hat{p} \geq 10$ and $n(1-\hat{p}) \geq 10$ (which means there are at least 10 successes and 10 failures in the sample)
Interval: $\hat{p} \pm z^*\sqrt{\hat{p}(1-\hat{p})/n}$
where $-z^*$ is the $(100\text{-}C)/2^{\text{th}}$ percentile of the standard normal distribution.

In Minitab:
Choose Stat > Basic Statistics > One Proportion
- Click the summarized data button and enter the value of n in the "number of trials" box and the observed number of successes (may need to multiply $n\hat{p}$ and round to the nearest integer) in the "number of events" box.
- Click Options and check the box next to "use test and interval based on normal distribution." (Otherwise Minitab produces the exact binomial p-value and confidence interval.)

For a test of significance:
- Click the Options button and specify the appropriate direction of the alternative hypothesis
For a confidence interval:
- Click the Options button and specify the appropriate confidence level
- Specify the alternative as "not equal"
Note: if you are just calculating a confidence interval, it does not matter what value you use for the test proportion and you will ignore that portion of the output. Click OK twice.

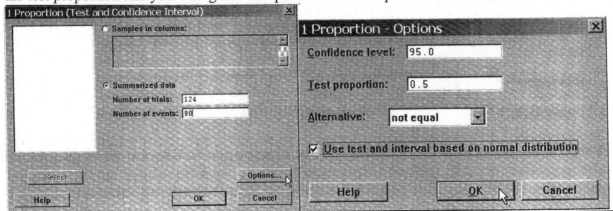

Values of the critical value z* for some common choices of confidence level appear below:

Confidence level	90%	95%	99%	99.9%
Critical value z*	1.645	1.960	2.576	3.291

Practice:

4-14) NCAA Gambling

A 2004 survey of 12,651 male NCAA athletes found 63.4% of Division I athletes participated in any gambling behavior (this includes playing cards, betting on games of skill, buying lottery tickets, betting on sports, inc.).
(a) Define the parameter of interest in this study.
(b) Do these data provide statistically significant evidence that more than 60% of Division I male athletes participate in any gambling behavior? Justify your answer (using all the steps of a test of significance), and explain your reasoning.
(c) Calculate and interpret a 95% confidence interval for the proportion of all Division I male athletes that participate in any gambling behavior.

4-15) More on Kissing

Reconsider Investigation 3-13.
(a) Determine the critical value z* for a 92% confidence interval.
(b) Calculate by hand a 90% confidence interval for π in the kissing study.
(c) How does the half-width of this interval compare to that of the 95% and 99% confidence intervals? Explain why this makes sense.
(d) Determine the *mid-point* of this interval. How does it compare to the mid-points of the 95% and 99% confidence intervals? Explain why this makes sense.

4-16) Margin of Error

Suggest two different ways for reducing the *margin of error* in a confidence interval for a population proportion π.

4-17) Sample Size Determination

Suppose you want to estimate the proportion of students at a very large university who are nearsighted. The prevalence of nearsightedness in the general population is 45%. Using this as a preliminary guess of π, how many students would need to be included in a random sample if you wanted the margin-of-error of a 95% confidence interval to be less than or equal to 1%?

Investigation 4-10: Reese's Pieces (Applet Exploration)

Previously you have made interpretations of a confidence interval as the set of plausible values for the population parameter (assuming the sample is representative of the population). In this investigation, you will use simulation to explore a more detailed interpretation of the *confidence level*. You will also investigate the effects that factors such as sample size and confidence level have on the confidence interval and the "coverage rate."

Earlier we assumed that 45% of all Reese's Pieces candies are orange.
(a) Using the sample proportion of orange candies that you found in Investigation 4-6, calculate a one-sample z-interval for the probability of obtaining an orange candy from this process.

(b) Did everyone in your class obtain the same confidence interval?

(c) Did everyone in your class obtain an interval that captured .45? Is this expected?

An important observation here is that confidence intervals vary from sample to sample. What you may find frustrating is that in a real study, you will have no way of knowing whether or not your interval does capture the parameter! In fact, here, the .45 value may not even be the correct value of π. In the following you will simulate a larger number of intervals where you can set the value of π to see how often these "random intervals" succeed in capturing π.

(d) Open the "Simulating Confidence Intervals" applet.
Set the value of π to be .45 and n to be 25.

Click Sample.

Click on the interval to see the sample proportion obtained (at the bottom). Use this sample proportion to verify the calculation of the endpoints of the interval displayed:

Simulating Confidence Intervals

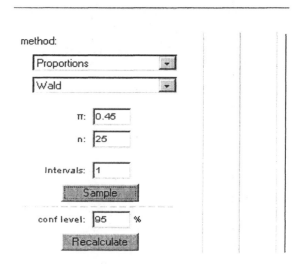

method:

| Proportions |
| Wald |

π: 0.45

n: 25

Intervals: 1

Sample

conf level: 95 %

Recalculate

Does this interval capture .45?

(e) Click Sample again. Did you get the same interval? Does this interval capture .45?

(f) Did the value of the *population* proportion of orange candies change between your two samples? How is this value represented on the graph?

To investigate what happens in the long-run, we will take many more random samples and construct a confidence interval for π from each.

(g) Change the number of intervals from 1 to 198 and click Sample. What percentage of the 200 intervals captures .45? (*Hint*: See the "Running Total." We will refer to this as the *coverage rate*.)

(h) Change the number of intervals from 198 to 200 and click Sample repeatedly until 1000 intervals have been created. From the running total, what percentage of these 1000 intervals captures .45? Is this coverage rate close to 95%?

(i) Click the Sort button. What do the red intervals (the ones that fail to capture the actual value of the population proportion) have in common?

(j) Predict how the intervals will change and how the coverage rate will change if you were to decrease the confidence level from 95% to 90%.

(k) In the applet, change the confidence level from 95% to 90% and click Recalculate. How did the individual intervals change? Continue to press Sample until 1000 90% confidence intervals have been created. What percentage of these 1000 intervals captures .45?

(l) Now recall that in an actual study, you only get one sample and therefore one interval. Would you have any way of knowing for sure if your interval is one of the intervals that succeeds or one that fails? Explain.

> These simulations reveal that if we take repeated random samples from a population with parameter π and calculate a C% confidence interval each time, then in the long-run roughly C% of the resulting intervals succeed in capturing π inside the interval. Thus, when we gather the actual sample data and calculate a *single* interval, we say that we are "C% confident" that this interval succeeds in capturing the actual value of the population proportion π. The basis for our confidence is knowing that the procedure would generate a successful interval in C% of all possible samples in the long run.

(m) Suppose you had 1000 confidence intervals for π. If you randomly select one interval, what is the probability that you will select an interval that captures π?

(n) Why is it technically incorrect to say "If I have calculated a 95% confidence interval for π, there is a .95 probability that π will be contained in the interval."?

Discussion: There is a very subtle distinction here. As worded, the above statement is making a probability statement about π and one specific interval. But our long-run relative frequency interpretation of probability talks about repeating the same random process many times and seeing how often certain outcomes occur. It does not make sense to make a probability statement about π, because π does not change. In other words, the population parameter π is not a random variable. Instead, we are considering repeated random samples from the same population. What *does* vary from sample to sample is the sample *proportion* and therefore the confidence *interval* that is constructed around that sample proportion. Technically, we could make a probability statement about this *method* capturing π. However, once you calculate an interval, it is no longer random (it either contains π or it does not, we just don't happen to know which). So we can talk about the probability that this method will produce an interval that contains the parameter, but we will not refer to the probability that the parameter is in a calculated interval or to the probability that a calculated interval contains the parameter. For these reasons, statisticians adopted the word "confidence" to describe the reliability of the method for producing a particular interval that succeeds in capturing the parameter of interest. Thus, before we collect our sample data, there is a .95 probability that if we construct a 95% confidence interval it will capture the population parameter. That is, this method "succeeds" for about 95% of intervals constructed from repeated random samples from this population. In reality, we obtain just one interval and once we determine that interval we say that we are 95% confident that this interval captures the population parameter.

EFFECTS OF SAMPLE SIZE, POPULATION PROPORTION

(o) Conjecture how the intervals and the coverage rate will change if we increase the sample size from $n=25$ to $n=75$ candies.

(p) Make this change in the applet and click Sample. Comment on how the intervals change. Is the coverage rate still approximately 95%? Were your conjectures correct? Explain.

(q) Conjecture how the intervals and the coverage rate will change if we increase the population proportion from $\pi = .45$ to $\pi = .65$.

(r) Make this change in the applet and click Sample. Comment on how the intervals change. Is the coverage rate still approximately 95%? Were your conjectures correct? Explain.

Investigation 4-11: Good News First or Bad News First

The Wald Interval is just one possible method for constructing a confidence interval for a population proportion. For example, you earlier studied a procedure based on the binomial distribution rather than the normal approximation, but it tends to be "conservative" (producing wider intervals than necessary for the desired coverage rate). The Wald procedure produces an approximate interval and requires a large enough sample size ($n\hat{p}$ and $n(1-\hat{p}) \geq 10$) to be valid. Researchers continue to propose other methods that will still achieve the stated 95% confidence level but work for small samples or tend to produce shorter (more precise) intervals. In this investigation you will explore and apply one of these procedures.

Suppose we take a random sample of students from your school and ask which they prefer to hear first—good news or bad news.

(a) Gather sample data on this question from your class. Then produce a bar graph to summarize the sample results. Do the students in your class prefer one option over the other?

(b) Define the population parameter of interest and calculate a 95% confidence interval for the proportion of students in your school who prefer to hear bad news first. Write a one-sentence summary of what this confidence interval says.

(c) Was the confidence interval in (b) a reasonable calculation? Are the conditions necessary for this procedure to be valid met?

ALTERNATIVE CONFIDENCE INTERVAL METHOD

Open the "Simulating Confidence Intervals" applet. Specify $\pi = .10$, $n = 30$, intervals = 200 and confidence = 95%. Generate 1000 intervals (200 at a time).

(d) What percentage of these 1000 intervals succeeded in capturing $\pi = .10$? Is this close to the claimed 95% confidence level?

Discussion: When the "technical conditions" for the Wald procedure are not met, the observed confidence level may be drastically below the stated confidence level. An alternative confidence interval procedure is to add two successes and two failures to the sample data before applying the above method, that is, using $\tilde{p} = (X+2)/(n+4)$ instead of $\hat{p}$ as the point estimate This is often referred to as the *Wilson estimator* (Wilson, 1927). This point estimate will also replace $\hat{p}$ in the calculation of the standard error: $\sqrt{\tilde{p}(1-\tilde{p})/(n+4)}$.

Adjusted Wald 95% confidence interval for π:

A 95% confidence interval procedure for π: $\tilde{p} \pm 1.96\sqrt{\tilde{p}(1-\tilde{p})/(n+4)}$ where $\tilde{p} = (X+2)/(n+4)$.
When n is small or moderate, this procedure is more likely to capture the population proportion close to 95% of the time.
Note: You can "fool" Minitab into doing this calculation for you by increasing the number of successes by two and the sample size by four.

Note: For more general confidence levels, using the appropriate z^* value, researchers have recommended using $\tilde{p} = (X+.5z^{*2})/(n+z^{*2})$ and $\tilde{n} = n+z^{*2}$ so the interval becomes

$$\tilde{p} \pm z^*\sqrt{\tilde{p}(1-\tilde{p})/\tilde{n}} \, .$$

This gives us a procedure we can apply even when the $n\hat{p} \geq 10$ and $n(1-\hat{p}) \geq 10$ conditions are not met. Still, we will probably want n to be at least 5 and we will still require the sample to be randomly selected from the population of interest.

(e) In the applet, use the pull-down menu to change the method from "Wald" to "Adjusted Wald." Generate 1000 intervals and comment on how the coverage rate compares to (d).

(f) Return to your class sample results on the "good and bad news" question. Use the adjusted Wald interval to estimate the population proportion who prefer to hear bad news first. Is this interval very different from the Wald interval that you calculated in (b)?

(g) Summarize the results of this study, referring to both your numerical and graphical summaries of the sample and your 95% confidence interval for the population proportion.

(h) Do you feel comfortable generalizing the results from this sample to all students at your school? Explain.

Discussion: You should have found that the adjusted Wald intervals are better at obtaining the nominal 95% confidence level than the Wald intervals. When the conditions for the normal approximation are indeed met, these procedures are similar. However even when $n\pi<10$ or $n(1-\pi)<10$, the adjusted Wald procedure still "works" (comes close to achieving the claimed confidence level). One lesson here is that checking technical conditions is very important, for you found that the Wald procedure can have a much lower coverage rate than claimed when its conditions are not met. Both procedures still rely on having a random sample to begin with.

Study Conclusions: If the entire confidence interval falls above .5, we have statistically significant evidence that students in your school prefer to hear bad news first *as long as the sample was representative of all students in your school on this issue*. In using the results from your class, we must have caution in generalizing these results to the larger population. What population are you willing to consider your class to be representative of? Do you have reason to suspect that statistics students this term will have a different preference for bad news than other students? What other considerations should be taken in order for the Bernoulli model to be valid here?

Practice:

4-18) Good News First?
Suppose we want to test whether students at your school prefer to hear bad news first.
(a) State the appropriate null and alternative hypotheses.
(b) What can you say about the p-value based on the calculations you performed in the previous investigation?

4-19) Seat Belt Usage
Every year since 1983 the Harris Poll has measured and reported the key lifestyle characteristics and behaviors which are known to have a major impact on health, disease, injury and life expectancy. One such survey was conducted by telephone with a nationwide sample of 1,005 adults between March 11 and 16, 1998. In this survey, 77% of adults claimed to always wear seat belts when they are in the front seat of a car.
(a) Define the relevant population parameter in words.
(b) Use this sample data to construct a 95% confidence interval for π using both the Wald and the adjusted Wald methods. How do the intervals compare?
(c) Explain what is meant by the phrase "95% confidence" for these procedures.
(d) Suggest a reason we might be skeptical that this confidence interval captures the proportion of adults who use a seat belt while in the front seat of a car. [*Hint*: Think about possible nonsampling errors.]

4-20) Wald vs. Adjusted Wald
Use the "Simulating Confidence Intervals" applet to compare the performance of the Wald and adjusted Wald confidence interval procedures for the following values of n and π. Use 95% confidence.
(a) $\pi = .15, n = 10$
(b) $\pi = .10, n = 100$
(c) $\pi = .50, n = 20$
(d) $\pi = .90, n = 50$
In each case report the percentage of intervals that successfully capture the parameter value for each procedure, using at least 1000 samples. Also write a paragraph summarizing your results.

4-21) *Literary Digest* (cont.)
Recall from chapter 3 that the *Literary Digest* conducted a large sample survey of 2.4 million people and found 57% indicating they would vote for Alf Landon in the upcoming 1936 election.
(a) Use Minitab to produce a 99.9% confidence interval for the proportion of all voters who were planning to vote for Alf Landon based on this sample data.
(b) Explain why the width of the interval is so narrow.
(c) Explain why the results of the interval provided such an erroneous estimate (the actual value of π turned out to be .37, as Landon received only 37% of the vote in the election).

SECTION 4-4: DISTRIBUTIONS OF SAMPLE MEANS

In this section we turn our attention from the sampling distribution of a sample *proportion* to the sampling distribution of a sample *mean*. In other words, we shift back to working with a quantitative rather than a categorical variable. You first learned about sampling variability of a sample mean in chapter 3 when you took samples of words from the Gettysburg Address and saw that the sample mean word length varied from sample to sample. Now you will investigate conditions under which the normal distribution is a reasonable model for describing the sampling distribution of a sample mean. You will see many similarities to the previous section, with one key change. You will also learn a new significance test and confidence interval procedure for a population mean.

Investigation 4-12: Scottish Militiamen and American Moms (Minitab Exploration)

The file `militiamen.mtw` contains chest measurements (in inches) for 5738 Scottish militiamen in the early 19th century. We will treat these observations as our population.

(a) What are the observational units and the variable of interest here? Is this variable quantitative or categorical?

Observational units:

Variable: Type:

(b) Use Minitab to create numerical and graphical summaries of this population and sketch the graph below. Describe the shape, center, and spread of this population distribution. Record the mean and standard deviation of the chest measurements. Are these values parameters or statistics? Explain. What symbols do we use to refer to them?

Shape:

Mean: Standard deviation:

Parameters or statistics?   Symbols:

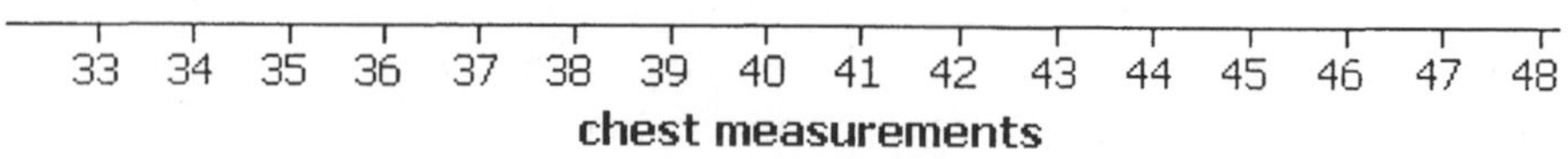

```
  33  34  35  36  37  38  39  40  41  42  43  44  45  46  47  48
                        chest measurements
```

(c) Use Minitab to select a random sample of 5 men from this population:

```
MTB> sample 5 c1 c2
MTB> name c2 'sample'
```

Produce numerical summaries and a *dotplot* of this sample, and describe the distribution. Record the mean and standard deviation; are they parameters or statistics? What symbols do we use to refer to them?

Shape:

Mean: Standard deviation:

Parameters or statistics?   Symbols:

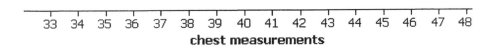

chest measurements

Discussion: While it is difficult to judge with only 5 observations, you should see that individual samples mimic the behavior of the population, but that sampling variability is alive and well. Your sample mean chest measurement $\bar{x}$ should be in the ballpark of the population mean ($\mu=39.832$), but of course different random samples produce different observations and therefore different sample means.

You will now investigate whether taking repeated random samples results in a predictable, long-run pattern of variation in sample means. In particular, you will take repeated samples from this normally distributed population to judge whether it is reasonable to model the sampling distribution of the sample statistic using a normal probability model. Then you will consider other possible population models/shapes and will develop criteria for determining when this normal model for sample means will be appropriate. Once you know these criteria, you will know in which situations you can apply the normal probability model to the distribution of sample means, even when you don't know the exact population distribution.

<u>Population 1</u>
(d) Keep the dotplot window open. Create the following Minitab macro to generate random samples from this population of chest measurements and calculate the sample mean and standard deviation for each sample:

```
sample 5 c1 c2
let c3(k1)=mean(c2)        #stores the sample mean, x̄
let c4(k1)=std(c2)         #stores the sample standard deviation, s
let k1=k1+1
```

In the session window, initialize k1 (`let k1=1`) and execute the macro once to make sure it is working (File > Other Files > Run an Exec). Then execute the macro 999 more times. For each sample, the sample mean is stored in C3 and the sample standard deviation is stored in C4. You should see the different samples appear in the dotplot window. Once you have a sense for how the values change from sample to sample, you can close this window (click the X in the upper right corner).

When the macro is done running, name C3 'xbar' and C4 's'. Produce numerical and graphical summaries of the distribution of sample means. What are the observational units and variable in this distribution? Describe the shape of this empirical sampling distribution of sample means in C3, and also report the mean and standard deviation of the 1000 sample means. How does this sampling distribution compare to the population distribution?

Observational units: variable:

Shape:

Mean: Standard deviation:

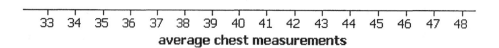

33 34 35 36 37 38 39 40 41 42 43 44 45 46 47 48
average chest measurements

Comparison to population shape, mean, and standard deviation:

Note: you have found empirical estimates for E($\overline{X}$) and SD($\overline{X}$), the expected value and standard deviation of the sample mean.

(e) Does the normal probability model appear to reasonably model the behavior of these sample means? *Hint*: Use your graph and a normal probability plot to judge.

Population 2

(f) The file agemom.mtw contains data gathered as part of the 1998 General Social Survey. Each mother in the sample reported what her age was when she gave birth to her first child. Column 1 contains data on 1199 American mothers. Open this worksheet (keep the Scottish chest measurements file open as well) and now consider these data as your population. Produce numerical and graphical summaries, and sketch and describe this population distribution.

Shape:

Mean: Standard deviation:

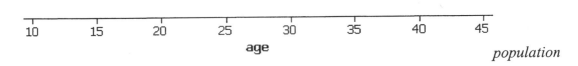

age   *population*

(g) Re-use the macro you created above (remembering to initialize your counter, k1=1) to generate an empirical sampling distribution of sample means (again for 1000 samples of size n=5) from this population of mothers' ages. Sketch and describe this distribution and use a probability plot to assess whether the normal model is appropriate (and if not, how violated). Also report the mean and standard deviation of the sample means.

Shape: Normal model?

Mean: Standard deviation:

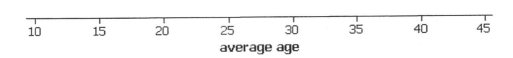

average age

(h) How do you think the sampling distribution for samples of size n=50 will compare?

(i) Now use the population of mothers' ages to generate an empirical sampling distribution of sample means for samples of size $n=50$. [Set k1 back to 1 and change the macro to select 1000 random samples of size $n=50$. Store the sample means and standard deviations in columns 5 and 6 this time.] Is this distribution reasonably modeled by the normal curve? Also report the mean and standard deviation of the sample means.

Shape: Normal model?

Mean: Standard deviation:

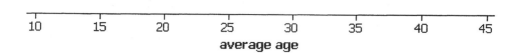

average age

(j) What are the two main differences between this sampling distribution (with $n=50$) and the sampling distribution in (g) with $n=5$? Do these changes agree with your conjecture in (h)?

(k) Complete the following table to summarize your observations:

Population	Sampling distribution of sample means		
	Shape	Center	Standard deviation
Normal $\mu=39.83$, $\sigma=2.05$, $n=5$			
Skewed right $\mu=22.52$, $\sigma=4.89$, $n=5$			
Skewed right $\mu=22.52$, $\sigma=4.89$, $n=50$			

[Focus on how the sampling distribution compares to the shape, center, and spread of the population. You should indicate whether the numerical summaries are in the ballpark, or quite different.]

Discussion: You should have observed that while the empirical sampling distribution of the sample mean chest measurements appeared normal with the small sample size ($n=5$), the empirical sampling distribution of the sample mean mother's age did not with $n=5$. However, with the larger sample size ($n=50$), the sample mean mother's age did follow a normal distribution, with less sampling variability. In each case, the mean of the sampling distribution was close to the mean of the population. The following theorem, probably the most important in all of statistics, summarizes and extends these observations in general.

Central Limit Theorem (CLT) for a Sample Mean:
Suppose that a random sample of size n is taken from a large population with mean μ and standard deviation σ. Then the sampling distribution of the sample mean $\overline{X}$ is reasonably modeled by a normal probability curve with mean equal to μ and standard deviation $\sigma/\sqrt{n}$, where the normal model applies exactly if the population distribution is normal, or approximately if the sample size is large. How large n needs to be in the second case depends on how far from normal the original population is, but a rough guideline is $n \geq 30$.

Note: The previous Central Limit Theorem about a sample proportion is actually a special case of this one. For a binary variable, if we think of the x_i outcomes as 0's (failures) and 1's (successes) and the sample proportion of successes $\hat{p}$ is equal to $\Sigma x_i / n$ and therefore is also a sample average.

(l) Calculate the value of the expression $\sigma/\sqrt{n}$ for the Scottish chest measurements and for both sample sizes with the mothers' ages. Record these in the table below, and verify that the standard deviations of your 1000 simulated sample means are close to this value in each case.

	$\sigma/\sqrt{n}$	Std. dev. of simulated sample means
Scottish chest measurements ($n=5$, $\sigma=2.050$)		
Mothers' ages ($n=5$, $\sigma=4.885$)		
Mothers' ages ($n=50$, $\sigma=4.885$)		

Probability Detour
The above results should convince you of the validity of the following probability rule.

When averaging independent and identically distributed random variables (with mean μ and standard deviation σ), the expected value of the sample mean is equal to the expected value of an individual observation: $E(\overline{X}) = E(X) = \mu$ and the standard deviation of the sample mean is equal to $SD(\overline{X}) = \sigma/\sqrt{n}$.

While these statements hold for independent observations (e.g., an SRS), the Central Limit Theorem further specifies when the shape will be normal or approximately normal.

APPLYING THE CENTRAL LIMIT THEORM

(m) Suppose we were to randomly select a sample of 5 Scottish chest measurements from this population with μ=39.83 in and σ=2.05in. Calculate the probability that the sample mean will exceed 41 inches, $P(\overline{X} > 41)$. [Include a sketch of the sampling distribution predicted by the Central Limit Theorem, labeling the mean and standard deviation and shading the area of interest. You may use Minitab or the applet to carry out the normal probability calculation. Confirm your answer with your simulation results.]

(n) Suppose we were to randomly select a sample of 50 mothers' ages from this population with μ=22.52 years and standard deviation σ=4.89 years. Approximate the probability that the sample mean age will exceed 24 years. [Include a well-labeled and scaled sketch.]

(o) Would it be reasonable to repeat (n) for a random sample of 5 mothers' ages? Explain.

(p) Would the probability in (o) be larger or smaller than the probability in (m)? Explain.

Discussion: For the Scottish militiamen, you should have found $P(\overline{X} > 41) = .095$ (and about 10% of your observations in C3 should exceed 41 inches). For the mothers' ages with n=50, $P(\overline{X} > 24) = .016$ (and about 1% of the observations in C5 exceed 24). We would not feel comfortable carrying out a normal probabability calculation with n=5 as the Central Limit Theorem does not apply. However, we know there will be more sampling variabilty and therefore a higher probabability of a sample mean exceeding 24 with the smaller sample size.

MINITAB EXPLORATION: CONFIDENCE INTERVAL FOR μ

(a) Based on these results (approximate normality with specified mean and standard deviation) suggest the form of a confidence interval to estimate the population mean μ.

Following the confidence interval form from the previous section, we could suggest estimating the population mean μ by the procedure:

$$\bar{x} \pm z^* \, \sigma/\sqrt{n}$$

as long as the conditions for the sampling distribution to be normal are met.

(b) Return to your 1000 simulated samples from the Scottish military men worksheet. For each sample, construct a 95% confidence interval of this form:

```
MTB> let c5='xbar'-1.96*std(c1)/sqrt(5)
MTB> let c6='xbar'+1.96*std(c1)/sqrt(5)
MTB> name c5 'lower' c6 'upper'
```

To determine how many of these intervals capture the population mean:

```
MTB> let c7=('lower'<mean(c1) & 'upper'>mean(c1))
MTB> tally c7
```

Is this percentage close to the nominal 95% confidence level?

Unfortunately, this expression involves the population standard deviation σ which will probably also be unknown to us. A natural adjustment is to estimate σ by the sample standard deviation, s:

$$\bar{x} \pm z^* \, s/\sqrt{n}$$

Def: The quantity $s/\sqrt{n}$ is referred to as the *standard error* of the sample mean, SE($\bar{X}$). This standard error provides an estimate of the standard deviation of the sampling distribution of the sample mean.

(c) For each simulated sample of Scottish chest measurements, construct a 95% interval of the above form (using the sample standard deviation each time):

```
MTB> let c8 = 'xbar' - 1.96*'s'/sqrt(5)
MTB> let c9 = 'xbar' + 1.96*'s'/sqrt(5)
MTB> name c8 'lower2' c9 'upper2'
```

and determine how many of these intervals capture the population mean:

```
MTB> let c10=('lower2'<mean(c1) & 'upper2'>mean(c1))
MTB> tally c10
```

Is this percentage close to the 95% confidence level?

Discussion: This procedure does not do well in achieving the claimed 95% coverage rate. The fix will be to multiply the standard error by a critical value bigger than 1.96 to compensate for the additional uncertainty introduced by estimating σ with s. The key to determining the new critical value is to compare the distributions of the following statistics that standardize the sample mean $\bar{x}$:

Statistic 1: $(\bar{x} - \mu)/(\sigma/\sqrt{n})$ Statistic 2: $(\bar{x} - \mu)/(s/\sqrt{n})$

Note that the only difference between these two statistics is the use of the sample standard deviation s in place of the population standard deviation σ in statistic 2.

(d) Use Minitab to calculate both statistics for each sample:

```
MTB> let c11=(c3-39.832)/(2.050/sqrt(5))
MTB> name c11 'stat1'
MTB> let c12=(c3-39.832)/(c4/sqrt(5))
MTB> name c12 'stat2'
```

Note: Minitab will produce missing values if a sample standard deviation in C4 equals 0, but you can proceed anyway.

Produce comparative boxplots (multiple Y's, simple) and write a few sentences contrasting the two sampling distributions.

(e) Produce normal probability plots for both statistics. Which one appears to follow a normal distribution?

(f) Describe how the sampling distribution of the other statistic deviates from a normal distribution.

Probability Detour: To model the sampling distribution of the statistic $(\bar{x} - \mu)/(s/\sqrt{n})$, we need a density curve with *heavier tails*. W. Gosset, a chemist turned statistician, showed in 1908, while working for the Guiness Breweries in Dublin, that a *t* probability curve with *n-1 degrees of freedom* provides a good model for the sampling distribution of this statistic if the population of observations follows a normal distribution. The probability density curve is described by the following function:

$$f(x) \propto \frac{1}{\sqrt{v\pi}(1 + x^2/v)^{(v+1)/2}} \text{ where } -\infty < x < \infty. .$$

An impressive function indeed! But you should notice that this function only depends on the parameter v, referred to as the *degrees of freedom.*

This symmetric distribution has *heavier tails* than the standard normal distribution. We get a different *t* distribution for each value of degrees of freedom. As the degrees of freedom increase, the *t* distribution approaches the standard normal distribution.

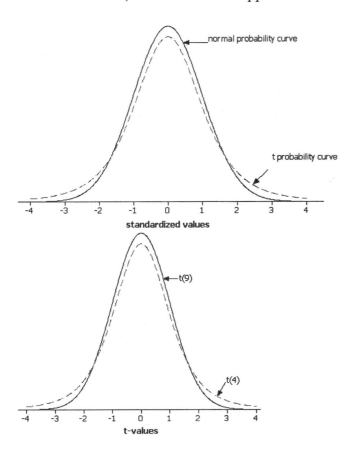

289

Thus we will want to use the t_{n-1} distribution to model the behavior of $(\bar{x} - \mu)/(s/\sqrt{n})$ instead of the standard normal distribution.

In particular, if we find $-t^*$ and t such that

$$P(-t^* \leq (\bar{x} - \mu)/(s/\sqrt{n}) \leq t^*) = C/100$$

then

$$P(\bar{x} - t^*s/\sqrt{n} \leq \mu \leq \bar{x} + t^*s/\sqrt{n}) = C/100$$

so $t^*s/\sqrt{n}$ is the margin of error and $\bar{x} \pm t^*s/\sqrt{n}$ is an approximate C% confidence interval for μ.

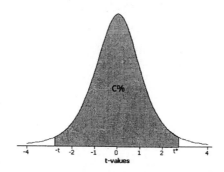

This leads to a confidence interval formula with the same general form as the previous section *sample estimate* ± (*critical value*) × (*SE estimate*), where the critical value now comes from the t_{n-1} distribution instead of the standard normal distribution. Keep in mind that the critical value again tells us how many standard errors we need to extend from the sample result depending on how confident we want to be. Our goal is to develop a C% confidence interval method that will capture the population parameter C% of the time in the long run.

(g) Find the t^* value corresponding to a 95% confidence level and $n = 5$ [*Hint*: Calc > Probability Distributions > t, using the "inverse cumulative probability" option, leave the "noncentrality parameter" set to 0, and enter 4 as the degrees of freedom, and .975 as the input constant.] How does this compare to the corresponding z^* value?

$t_4^* = z^* =$

Comparison:

(h) Recalculate the 1000 confidence intervals in (o) using this t^* value in place of 1.96. What percentage of the resulting intervals succeed in capturing the population mean $\mu = 39.832$?

(i) Does this method appear to be a 95% confidence interval procedure? Explain.

Discussion: These t-intervals will be slightly larger than if we used z^* as the critical value (since $t^* > z^*$). This is the price we have to pay for estimating the population standard deviation, but the adjustment of using the t critical value appears to be a valid way to obtain the desired confidence level.

An approximate **C% confidence interval for** μ can be calculated using the formula:

$$\bar{x} \pm t^*_{n-1}\, s/\sqrt{n}$$

where $-t^*$ is the $(100\text{-}C)/2^{\text{th}}$ percentile of the t distribution with n-1 degrees of freedom. This procedure will be considered valid fs we have a random sample from the population of interest and *either* a normal population or a sample size of at least 30. The procedure is *robust* in that the coverage rates are reasonably stable even for smaller sample sizes and slightly skewed distributions.

Minitab Detour: To calculate these t intervals in Minitab, choose Stat > Basic Statistics > 1 Sample t. You can either tell Minitab which column has your sample, or you can enter the summary statistics (n, $\bar{x}$, and s). To change the confidence level, choose the Options button.

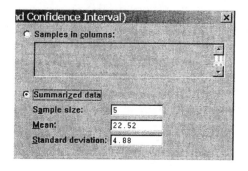

Practice:

4-22) Mothers' Ages (cont.)
Suppose you obtain a sample mean of 22.52 years and sample standard deviation 4.88 years and plan to construct a 95% confidence interval for μ, the population mean age of mothers at first birth.
(a) What additional information do you need to construct the confidence interval?
(b) Calculate a 95% confidence interval for $n=5$, $n=25$, and $n=100$.
 (c) How do the intervals in (b) compare? Explain why this makes sense.
(d) Suppose the sample standard deviation had been 9.75 years instead. Recalculate the 95% confidence interval for $n=25$. How does it compare to the interval with $s=4.88$? Explain why this makes sense.
(e) Calculate a 99% confidence interval for $n=25$. How does it compare to the interval with C=95. Explain why this makes sense.
(f) Suppose the sample mean had been 32.52, $n=25$, C=95%. How does the interval change? Explain why this makes sense.

4-23) Body Temperatures (cont.)
Recall the 1992 study on body temperatures (*JAMA*). The researchers cite problems with Carl Wunderlich's "axioms on clinical thermometry" and claim that the traditional value of 98.6 is out of date. Subjects were healthy men and women, aged 18-40 years, who were volunteers participating in Shigella vaccine trails conducted at the University of Maryland Center for Vaccine Development, Baltimore. The researchers took the subjects' normal oral temperatures at several different times during three consecutive days using an electronic digital thermometer. The data in temps.mtw were constructed by Shoemaker (1996) to match the histograms and summary statistics given in the article as closely as possible, but ensuring an equal number of men and women (65).
(a) Produce numerical and graphical summaries to describe the distribution of body temperatures in this sample. Summarize what they reveal. Does the sample appear well modeled by a normal distribution?
(b) Check the conditions for the validity of the *t* confidence interval. Do you feel comfortable applying the procedure to these data? Explain.
(c) Define the population parameter of interest in this study.
(d) Calculate a 95% confidence interval for this parameter. Provide a one-sentence interpretation of this interval.
(e) Does this confidence interval provide evidence that the mean body temperature differs from 98.6 degrees Fahrenheit? Explain.

APPLET EXPLORATIONS

Exploration 1: Return to the Simulating Confidence Intervals applet. Change the "method" from Proportions to Means and start with "z with s." Specify 39.8 as the population mean, 10 as the population standard deviation, 5 as the sample size, 200 as the number of intervals, and 90% as the confidence level.

(a) Examine 1000 confidence intervals (200 at a time). Do approximately 90% of the intervals succeed in capturing μ?
(b) Change from "z with s" to "t" and repeat (a).
(c) Repeat (a) and (b) with a sample size of $n=25$. How do the intervals themselves change with this increase in sample size and roughly how does the observed confidence level compare to the stated confidence level?
(d) Press the Sort button. What types of $\bar{x}$ values lead to intervals that fail to capture μ?
(e) Are all of the intervals the same length? Why not?
(f) Write a paragraph explaining why the t procedures are preferred to "z with s".

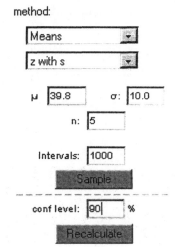

Exploration 2: Open the "Simulating t Confidence Intervals" applet Specify 39.8 as the population mean, 10 as the population standard deviation, 50 as the sample size, 200 as the number of intervals, and 90% as the confidence level. Click Sample. The dotplot in the upper right is the empirical sampling distribution of the sample means, the dotplot in the lower right is the distribution of the last sample.

(a) Describe each distribution. How are they similar? How are they different?
(b) Do approximately 90% of the intervals succeed in capturing μ? (Look at 1000 intervals.)
(c) Change the population distribution from Normal to Exponential and specify the value of μ to be 5 with a sample size of 50. Click Sample. Now how do the two distributions compare? Explain why each has the shape that it does. Do approximately 90% of the intervals succeed in capturing μ?
(d) Now change the sample size from 50 to 15 and repeat (b) and (c). Then change the sample size to 5 and repeat (b) and (c) again.
(e) Change the population distribution to Uniform, specifying the endpoints to be $a=5$ and $b=10$. Repeat (c) for sample sizes of $n=5$, $n=15$, and $n=50$.
(f) Explain what is meant by a "uniform" distribution.
(g) How well does the t procedure do in (e), even when the sample size is small? Suggest an explanation for why the t

confidence interval has better "coverage" with the uniform distribution than with the exponential distribution.

294

Investigation 4-13: Basketball Scoring

Prior to the 1999-2000 NBA season, rules were changed in an effort to increase scoring and make the game more exciting. Previously, the league averaged 183.2 points per game. Open the file NBApts99.mtw in Minitab. This file contains sample data on the total number of points scored for games between December 10 and 12, 1999.

(a) Use Minitab to construct a graphical display and numerical summaries of the sample and comment on the shape, center, and spread of the distribution.

(b) Define the population parameter of interest in this study. Then state the null and alternative hypotheses (in symbols and in words) for testing whether the average points per game has increased since the rule change.

Parameter:

H_0:

H_a:

(c) Suggest a way of measuring how far the observed sample mean is from the value conjectured for the population mean in (b), based on the sample data.

(d) Suppose we did have a random sample of games after the rule change and that the distribution of points scored in the population is normal. If the null hypothesis that the population mean equals 183.2 were true, then what distribution would the test statistic $t_0 = (\bar{x} - \mu_0)/(s/\sqrt{n})$ have? Sketch the sampling distribution below and label the horizontal axis.

(e) Are the conditions for the Central Limit Theorem of Sample Means suitably met to convince you that the sampling distribution of the sample means will be normal or approximately normal? Explain.

This test statistic has an exact t distribution when the population is normal and an approximate t distribution when the sample size is large.

Since the sample size is moderately small, this test statistic will be well modeled by a t distribution if the distribution of the population of points scored follows a normal distribution. Since we aren't actually able to examine the population distribution, we use the sample data in order to judge whether this is a reasonable assumption about the population distribution. If the sample is well modeled by a normal distribution, we will be willing to believe the population distribution is as well. Since the t-procedures are *robust*, as you saw in the previous investigation, we only refrain from using t-procedures if the sample size is small and the sample data have a substantial skew or extreme outliers.

(f) Use your graph in (a) and a normal probability plot to decide whether the sample data appear to be reasonably modeled by the normal distribution.

(g) In order for this test statistic to follow a t distribution, we also need the sample to be randomly selected from the population of interest here. Do you think this condition is met? Explain.

Even though this sample of games was collected in a three-day period and so is not a random sample, we will proceed for now to illustrate the calculations of a "one-sample t-test." The test statistic follows the same form as the "one-sample z-test" in that we are comparing the observed result (sample mean) to the hypothesized value, dividing by an estimate of the standard deviation of the observations (SE(sample mean)):

$$t_0 = \frac{\bar{x} - \mu_o}{s / \sqrt{n}}$$

But because we have to estimate σ with s, we will again compare this test statistic to the t distribution with n-1 degrees of freedom to calculate the p-value. As always, the direction of the p-value is determined by the alternative hypothesis statement, and small p-values constitute evidence against the null hypothesis.

It is always helpful to accompany your calculation of the p-value with a sketch. With the *t*-test, we will sketch the sampling distribution of the *t*-statistics rather than the sampling distribution of the sample means:

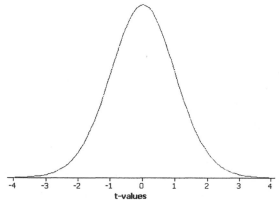

The shape of this distribution is always symmetric and centered at 0. The heaviness of the tails will be determined by the degrees of freedom (n-1). As the degrees of freedom increase, this distribution behaves more and more like the standard normal distribution.

(h) What is the value of this test statistic for the above NBA data ($\bar{x}$ = 195.88 points, s = 20.27 points)? Where does this value fall in the above distribution? Shade the area under the curve corresponding to values at least this extreme. Estimate the value of the p-value from your graph.

Observed *t*:   estimate of p-value:

(i) To calculate the p-value in Minitab, choose Calc > Probability Distribution > t, choose the Cumulative probability option, specify the appropriate degrees of freedom, and enter the observed test statistic value as the input constant. Does this probability represent the p-value?

(j) Would you reject or fail to reject the null hypothesis at the α = .05 significance level? State your conclusion in context.

This test produces a very small p-value (t=3.13, p-value=.005), providing very strong evidence that the population mean points per game in the new season exceeds 183.2. But like always, the test does not give any indication of *how many more* than 183.2 points per game were scored on average. Following up the test with a confidence interval enables us to estimate the magnitude of this difference.

(k) Use Minitab to determine the appropriate critical value for a 90% confidence interval. Then compute the interval. What does this interval indicate to you about the effect of the rule changes on the scoring average in the league? Explain. [Remember that the confidence interval specifies a set of *plausible* values for the parameter. Also consider the type of study conducted here.]

(l) Determine how many of the sample games have points that fall within the interval. [*Hints*: You might want to go through the values of C1 and count or look at your graphical display in (a). You can also create an indicator variable using the let command to determine whether each value falls above the lower bound and below the upper bound (see Investigation 4-12). Then use tally.] Is this close to 90% of the games in the sample? Explain why there is no reason for it to be.

This last question is meant to remind you that a confidence interval estimates the value of a population *parameter*, in this case the *mean* points per game in the NBA that season. A confidence interval makes no attempt to estimate the value of a particular observation, so there is no reason to expect 90% of the sample values (or 90% of the population values) to fall within the interval.

(m) If the sample size had been larger (say, 100), would you expect that about 90% of the games' point totals would fall within a 90% confidence interval for μ? Or would you expect an even smaller percentage of the games point totals to fall within that interval? Explain. [*Hint*: Think about how the interval will change based on the larger sample size.]

PREDICTION INTERVALS

In many situations we are more interested in estimating an *individual* value instead of the population mean.

(n) If we were to predict the number of points scored in a future game, what one number is your best estimate of that outcome (assuming the game is coming from the same population)?

(o) Roughly how far do you expect an individual observation to be from the population mean?

Def: To predict an individual value, we can calculate a *prediction interval* (PI). We construct the interval using the sample mean as our estimate, but we adjust the standard error to take into account the additional variability of an individual value away from the population mean:

$$\bar{x} \pm t^*_{n-1}\sqrt{s^2 + s^2/n} = \bar{x} \pm t^*_{n-1}\, s\,\sqrt{1 + 1/n}$$

This procedure is valid as long as the sample observations are randomly selected from a normal distributed population. Note that prediction intervals are *not* robust to violations from the normality condition even with large sample sizes..

(p) Calculate a 90% prediction interval based on this sample. Include a one-sentence summary of what this interval says.

(q) How does the width of this prediction interval in (p) compare to the width of the confidence interval for the population mean in (k)? Explain why this makes sense.

(r) What proportion of the sample results fall within this interval? Is this close to what you would expect? Explain.

Technology Detour

(s) In Minitab, choose Stat > Basic Statistics > 1-Sample t.

Click in the "Samples in columns" box.
Double click on "points" to enter that
column as the variable, and specify the
hypothesized value of μ in the "Test mean"
box. Click the Options button and use the
pull-down menu to specify the direction of
the direction of the alternative hypothesis
stated in (b). Click OK twice.

Verify your calculations of the test statistic
and p-value.

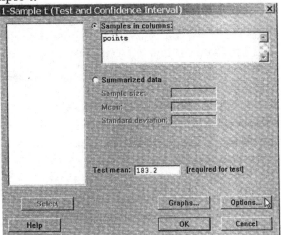

(t) Verify your calculation of the 90% confidence interval [You will need to change the
alternative back to "not equal to".

Applet Alternative:
Open the "Test of Significance Calculator" applet.
Use the pull-down menu to select "One mean."
Enter the hypothesized value, specify the direction
of the alternative, and enter the summary statistics.

Significance Test Calculations

Pressing the 95% CI button reports the 95%
confidence interval for μ.

(u) How does the 95% confidence interval
compare to the 90% confidence interval?

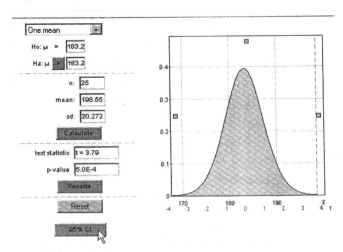

(v) Move the vertical blue line to the left until the
p-value is just below .05. What is the minimum
value of the test statistic for which you would
reject at the 5% level?

Study Conclusions: The points scored in these 25 games from the 1999-2000 season are fairly symmetric with mean $\bar{x}$ = 195.88 points and standard deviation s=20.272 points.

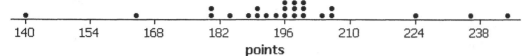

Most of the results are above the previous scoring average, before the rule change.

In fact, the sample data provide strong evidence ($t = 3.13$, p-value = .002) that the mean points per game in the season following the rule change, μ, is higher than the 183.2 mean value from the previous season. This conclusion stems from the p-value indicating that if the population mean for the new season were still 183.2, there's only a 2/1000 chance that random sampling variation alone would produce a sample mean points per games at least as large as the $\bar{x}$ = 195.88 value that was observed. A sample mean would only need to 1.71 standard errors above 183.2 to be statistically significant at the 5% level. This result was over 3 standard errors above the hypothesized value.

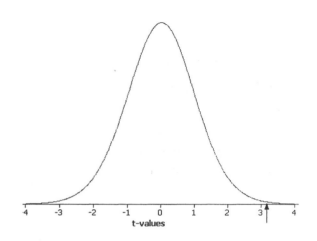

We are 90% confident that the value of μ is between 187.5 points and 202.2, and we are 90% confident that the points scored in a future game will be between 152.3 points and 239.5 points. While the sample size ($n = 25$) is not terribly large, the distribution of points scored appears to be reasonably symmetric, so that condition for the validity of the *t*-test appears to be satisfied. The random sampling condition is worrisome, though, because the data were not collected randomly but over one weekend early in the season. One could question whether that weekend's games are representative of the entire season; perhaps teams required more time to learn to adjust their defensive strategies to the rule changes. Because of this concern, the conclusion that scoring has increased on average should be taken with a grain of salt. Even if this was a random sample, it is not a randomized experiment, and we cannot draw a cause and effect conclusion that the increase in average scoring is due exclusively to the rule change.

Summary of One-Sample t Procedures for Mean

Test of H_0: $\mu = \mu_0$
Technical conditions: Random sample and either
$n \geq 30$ or normal population

Test statistic: $t_0 = (\bar{x} - \mu_0)/(s/\sqrt{n})$

p-value (with n-1 degrees of freedom):
      If H_a: $\mu > \mu_0$, the p-value is $P(T_{n-1} \geq t_0)$.
      If H_a: $\mu < \mu_0$, the p-value is $P(T_{n-1} \leq t_0)$.
      If H_a: $\mu \neq \mu_0$, the p-value is $2P(T_{n-1} \geq |t_0|)$

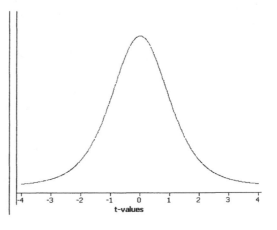

C% confidence interval for μ:
Technical conditions: Random sample and either $n \geq 30$ or normal population

Interval: $\bar{x} \pm t^*_{n-1} \, s/\sqrt{n}$
where $-t^*$ is the $(100-C)/2^{\text{th}}$ percentile from the t distribution with n-1 degrees of freedom.

In Minitab, choose Stat > Basic Statistics > 1-Sample t
- If the raw data are in a column, pass that column into the "Samples in columns" box
- If you only have the summarized data (sample mean $\bar{x}$, sample standard deviation s and sample size n), select "Summarized data" and enter the information.

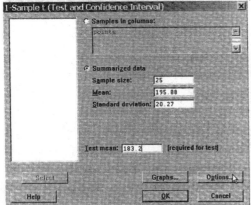
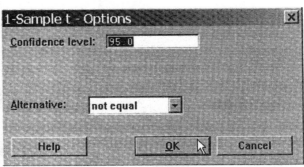

For a test of significance
- Specify the hypothesized value of μ and under Options specify the direction of the alternative.
For a confidence interval
- Make sure the alternative is set to "not equal" and specify the confidence level under Options.

C% prediction interval for an individual observation:
Technical conditions: Random sample and normal population

Interval: $\bar{x} \pm t^*_{n-1} \, s \sqrt{1+1/n}$
Minitab does not compute prediction intervals in this setting.

Note: The p-value and confidence interval methods are *robust* to minor departures from the normality condition but the prediction interval is not.

Investigation 4-14: Comparison Shopping (cont.)

A student group at Cal Poly carried out a study to compare the prices at two different local grocery stores. The inventory list for Scolari's (a privately-owned local store) was broken into sheets, each with 30 items. A number between 01 and 30 was randomly generated, 17, and the 17th item on each sheet was selected. Each student was then responsible for obtaining the price of that item at both Scolari's and Lucky's (which advertises itself as a discount grocery store) If the exact item was not available at both stores, the item was adjusted slightly (size or brand name) so that the identical item could be priced at both stores. Students at Cal Poly gathered prices for a set of 29 items. The data are available in `shopping99.mtw`.

(a) Define the observational units, population and sample in this study. Which store do you suspect will have lower prices?

(b) Was this a simple random sample, stratified sample, multistage cluster sample or systematic sample?

(c) Use Minitab to produce numerical and graphical summaries of the price distribution at each store. Identify any outlying values by the product name (Editor > Brush, Editor > Set ID Variables (C1)).

(d) Suggest a better way of comparing the two stores' prices on these 29 products.

Note that *pairing* items in this way is a form of "blocking" (see chapter 1). In fact, when there are just two observational units in each grouping or two responses for the same observational unit, the design is often called a *matched pairs design* (though the terms "blocking" and even "matched pairs" are usually reserved for experimental designs).

Here, by using the same product at both stores we control for the variability in prices from product to product, which improves the efficiency of the study by only comparing apples to apples and toothpaste to toothpaste. To analyze these paired data, we will examine the *differences* in the price for each pair.

(e) Use Minitab to calculate the price *differences* for these items (MTB> let c4=c2-c3, name this column "diffs") and produce numerical and graphical summaries for these differences. Does there appear to be evidence that one store tends to charge higher prices than the other? Explain.

(f) Identify any outliers by name. Are these the same unusual products you identified in (b)? Examine the data window—can you suggest an explanation for any of these outliers?

(g) Would it be appropriate to remove some of these outliers from the data set? What is the justification? Remove all products from the data set that meet this same justification. (Click on the row number and press Delete.) Now what is the size of your sample?

(h) Reproduce numerical and graphical summaries for this sample. Does there appear to be evidence that one store tends to charge higher prices than the other? Explain.

(i) Let μ represent the mean price difference of all items common to these two stores. State a null and alternative hypothesis about μ that reflects your initial conjecture about the two stores.

(h) Carry out a one-sample *t*-test on the price differences (often called a *paired t-test*) by sketching the sampling distribution of the test statistic, calculating the observed test statistic and p-value. Do you reject or fail to reject the null hypothesis at the 10% level? What recommendation would you give to a shopper choosing between the two stores?

(i) Provide and interpret a 90% confidence interval for the mean price difference between these two stores.

Study Conclusions: An important first step in data analysis is always to explore your data! With these data, we found some unusual observations and upon further investigation realized that some data values had been recorded in error. In this case, we would be justified in removing these observations from the data file. After cleaning the data, we found that the price differences were slightly skewed to the left with a few outliers (flour, toothpaste, and frozen yogurt). The average price difference (Lucky's − Scolari's) was − $.118, with a standard deviation of $.359. The median price difference was $0. So the sample did not show a strong tendency for one store to have lower prices, the sample mean was in the conjectured direction. A one-sided *paired t-test* found that the mean price difference between these two stores was significantly greater than 0 (*t*-value = − 1.74, p-value = .046) at the 10% level of significance, and even (barely) at the 5% level. A 90% confidence interval for the mean price difference was (-0.233701, − 0.002728). We are 90% confident that, on average, items at Scolari's cost between .2 cents and 23 cents more than items at Lucky's. This seems like a small savings but could become practically significant for a very large shopping trip. We feel comfortable generalizing these conclusions to the population of all products common to the two stores since the data were randomly selected using a probability method.

Practice:

4-24) Schizophrenic Twins (cont.)

Scientists have long been interested in whether there are physiological indicators of diseases such as schizophrenia. In a 1990 study by Suddath el. al., reported in Ramsey and Schafer (2002), researchers used magnetic resonance imaging to measure the volumes of various regions of the brain for a sample of 15 monozygotic twins, where one twin was affected by schizophrenia and the other not ("unaffected"). The twins were found in a search through the United States and Canada, the ages ranged from 25 to 44 years with 8 male and 7 female pairs. The data (in cubic centimeters) for the left hippocampus region of the brain appear below and in the Minitab worksheet `Hippocampus.mtw`:

(a) Calculate the difference in hippocampus volumes for each pair of twins.

(b) Calculate and interpret a 95% confidence interval for the mean volume difference using the t distribution. Also comment on the validity of this procedure.

(c) Based on this confidence interval, is there is statistically significant evidence that the mean difference in left hippocampus volumes is different from zero?

4-25) Body Temperatures (cont.)

(a) State the null and alternative hypothesis to see if average temperature of a healthy adult now different from 98.6 degrees Fahrenheit. Make sure you define the population parameter in words.

(b) Use the one-sample t-test to decide if there is statistically significant evidence that the mean body temperature differs from 98.6 degrees Fahrenheit (`temps.mtw`). Make sure you sketch the sampling distribution, check the technical conditions, report the test statistic and p-value, and state your conclusion in context.

(c) Explain what is meant by a Type I and a Type II error in this comparison.

(d) Calculate a 95% prediction interval for an individual's body temperature. At what temperatures should someone become concerned about their health?

4-26) Confidence vs. Prediction Intervals

(a) Examine the expression for a confidence interval for a population mean μ. What happens to the half-width of the interval as the sample size n increases? Describe its limiting behavior.

(b) Examine the expression for a prediction interval for an individual observation. What happens to the half-width of the interval as the sample size n increases? Describe its limiting behavior.

(c) Explain why the differences in your answers to (a) and (b) make sense.

SECTION 4-5: BOOTSTRAPPING

In the previous sections you explored whether the normal probability model was an appropriate model for the sampling distribution of sample proportions and for the sampling distribution of sample means. For binary data, if the normal model is not appropriate, we could use the binomial distribution as in chapter 3. For quantitative data, we did not often have an "exact" probability model to fall back on. In this section you will explore an alternative for when the distribution of sample means is not well modeled by a normal distribution, as well as for modeling the sampling distribution of other sample statistics where theoretical probability models have not been developed. This method, *bootstrapping*, has grown in popularity in recent years with the increasing power of computers. We will first introduce bootstrapping in a case where we know "the answer," the Gettysburg Address data, to illustrate how the method works. Once you understand the method, you will be able to use it when you don't have access to the entire population nor know the theoretical sampling distribution of the statistic, but you still want to make inferential statements about a population parameter.

Investigation 4-15: Sampling Words (Minitab Exploration)

Recall the population of words in the Gettysburg address with mean $\mu = 4.29$ letters and standard deviation $\sigma = 2.12$ letters. Suppose we take a random sample of $n=10$ words from this population. Let $\overline{X}$ represent the sample mean.

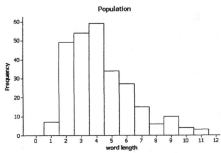

(a) Calculate the expected value of $\overline{X}$ and the standard deviation of $\overline{X}$.

(b) According to the Central Limit Theorem for a sample mean, do you expect this sampling distribution to be well modeled by a normal distribution? Explain.

In working with the sample mean $\overline{X}$, probability theory tells us the expected value and the standard deviation of $\overline{X}$, and the Central Limit Theorem tells us when the distribution will be approximately normal. According to the Central Limit Theorem, since the population of word lengths was skewed and the sample size is small, the sampling distribution of $\overline{X}$ might not be well modeled by a normal distribution. In this example, since we actually have access to the population of word lengths, we can use simulation to approximate the sampling distribution of the sample mean to confirm this. Below is an empirical sampling distribution of sample mean word lengths and their corresponding *t*-statistics based on 1000 samples of size *n*=10 from this population.

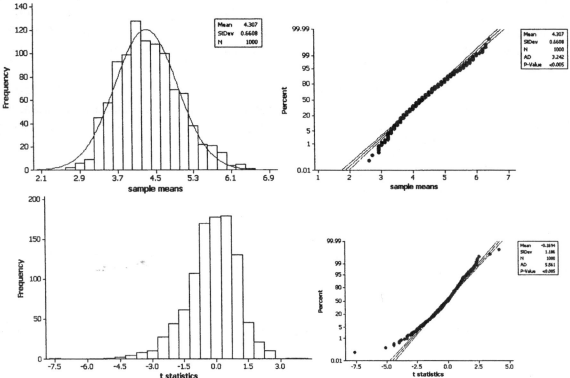

As we suspected, the empirical sampling distribution is centered near μ, the empirical standard deviation is approximately $\sigma/\sqrt{n} = 2.123/\sqrt{10} = .67$, but the distribution still has some skewness to it, and the standardized values, $(\overline{x} - \mu)/(s/\sqrt{n})$, are not symmetric. Consequently, *t*-confidence intervals may not be valid (though you saw earlier that the *t* intervals are fairly robust to the normality condition. Indeed, if we were to calculate 95% confidence intervals from each simulated sample, then about 95% of them would captured μ). However, we are not always so lucky that these *t* procedures work when the distribution of *t* statistics is not symmetric.

In general, there are two possible complications—the sampling distribution of $\overline{X}$ is not well modeled by a normal or *t* distribution (population distribution is skewed and sample size is small) or we want to work with a statistic other than the sample mean for which we don't know how to determine the standard deviation of its sampling distribution. In order to make inferences about the population from a random sample in these cases, we need other methods for estimating the amount of sampling variability in the sample statistic and the behavior of its sampling distribution. The technique of "bootstrapping" has recently been proposed as a method for doing this.

Def: A *bootstrap sample* resamples the data from the existing sample, drawing the same number of observations, but *with replacement*.

The reasoning behind this technique is that if the sample has been randomly selected, it should be representative of the population. Thus, it should give us information about the population and about samples drawn from that population. In other words, rather than assume a particular probability model for the population, bootstrapping assumes that the population looks just like the sample, replicated infinitely many times. By sampling with replacement from the original sample, and calculating the statistics of interest for each bootstrap sample, we gain information about the shape and spread of the sampling distribution of the statistic of interest.

(c) Open the `gettysburgSample.mtw` worksheet. Column 1 contains a random sample of 10 words and column 2 specifies their lengths. Determine the mean and standard deviation of this sample, as well as a 95% *t*-confidence interval for the population mean based on this sample.

sample mean: sample standard deviation:

95% *t*-interval

(d) To take a bootstrap sample from this population, we will sample *with replacement*:
```
MTB> sample 10 (c1 c2) (c3 c4);
SUBC> replace.
MTB> mean c4
```
What bootstrap sample mean did you obtain? Does this equal the mean of the original sample?

(e) Repeat (d), did you obtain the same bootstrap sample mean this time?

This bootstrap process randomly selects 10 words from the sample. Since we are sampling with replacement, it is likely that the same word will occur more than once in our bootstrap sample. The key is that the probability that a word length is selected for a bootstrap sample is equal to the proportion of words of that length in the sample and assumed to be the same as in the population. The assumption is that this "infinite" version of the sample mirrors the behavior of the population. We may have obtained an unlucky sample to begin with, in which case bootstrapping does not magically give us more correct information about the population, but this is a "risk" that we always take when we make inferences from the sample to the larger population. However, by using probability sampling methods to select the sample we minimize this risk.

(f) Create a Minitab macro to replicate this bootstrapping process. [*Hint*: Use `let c5(k1)=mean(c4)` to store the bootstrap sample means.] Remember to initialize the counter, and then execute the macro 1000 times to draw 1000 bootstrap samples. Create a (well-labeled) graph of this bootstrap distribution.

(g) What are the mean and standard deviation of the bootstrap sample means? Does the standard deviation concur with the value predicted in (a)?

Discussion: The empirical distribution of the bootstrap sample means should have the same standard deviation as the sampling distribution of $\bar{X}$ ($\sigma/\sqrt{n} = .671$) but will be centered at the original sample mean (4.8) instead of the population mean (4.29), as illustrated in the histogram below. Your empirical bootstrap distribution may not match these values exactly, but would get closer to them if you took more bootstrap samples. This exercise should help convince you that bootstrapping does provide another way to estimate this standard deviation.

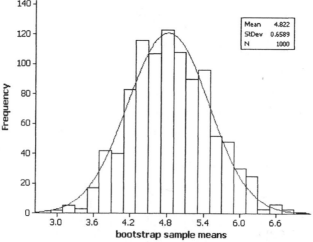

So bootstrapping doesn't change our estimate of μ, but it does give us a measure of how far our sample statistic might reasonably fall from μ. A very rough 95% confidence interval for μ

would then be to go 2 times this estimated standard error on either side of the sample statistic. A slightly rough bootstrap confidence interval would replace the multiplier of 2 with t^*_{n-1} for 95% confidence. Since the standard deviation of the bootstrap sample means is close to $\sigma/\sqrt{n}$, this produces an interval very similar to the t-interval.

(h) Use the sample mean from (c) and your estimate of the standard deviation from (g) to create this rough bootstrap confidence interval for μ. How does it compare to the t confidence interval from (c)?

To determine a bootstrap confidence interval for a population parameter θ, apply this same logic:
1. Generate a bootstrap distribution by sampling with replacement form the original sample and calculating the sample statistic of interest from each bootstrap sample.
2. Calculate the standard deviation of these bootstrap sample statistics as an estimate of the standard deviation of the sampling distribution of the statistic.
3. Create a bootstrap confidence interval by taking:
original sample statistic $\pm\ t^*_{n-1} \times$ estimate of std dev of statistic

---

FURTHER BOOTSTRAP EXPLORATION

For most statistics, the bootstrap distribution also provides information about the shape of the sampling distribution. Since the bootstrap distribution is slightly skewed, this often indicates skewness in the sampling distribution as well. When the sampling distribution is skewed, we might prefer a confidence interval that reflects that skewness rather than use a symmetric confidence interval as above.

The following notation will help us to derive a general bootstrap confidence interval procedure:
- θ, the population parameter
- $\hat{\theta}$, the sample estimate of θ
- $\hat{\theta}^*$, the ith bootstrap sample estimate of θ.

To create a 95% confidence interval for θ based on the sample estimate $\hat{\theta}$, we need to determine the distance that we plausibly expect $\hat{\theta}$ to fall from θ (at the 5% level). If we knew the critical values c_1 and c_2 such that
$$P(c_1 \leq (\hat{\theta} - \theta) \leq c_2) = .95$$
then we could rearrange the inequalities as follows:
$$P(\hat{\theta} - c_2 \leq \theta \leq \hat{\theta} - c_1) = .95,$$
to produce a 95% confidence interval for θ.

However, we don't know the theoretical sampling distribution of $\hat{\theta}$ (or of $\hat{\theta} - \theta$), so we need another way of determining the values of c_1 and c_2. Bootstrapping will enable us to determine these values based on the following key result.

Key Result: The distribution of $\hat{\theta} - \theta$ is closely approximated by the distribution of $\hat{\theta}^* - \hat{\theta}$.

In particular, the percentiles of these two distributions match. So the 97.5[th] percentile of the distribution of $\hat{\theta} - \theta$ is the same as the 97.5[th] percentile of the distribution of $\hat{\theta}^* - \hat{\theta}$. Fortunately, we can find the 97.5[th] percentile of $\hat{\theta}^*$ from the bootstrap sampling distribution and then subtract $\hat{\theta}$ to find the 97.5[th] percentile of $\hat{\theta}^* - \hat{\theta}$.

(i) Return to the bootstrap samples from the Gettysburg Address that you generated above. To find the 97.5[th] percentile of the bootstrap sample means stored in C5, $\hat{\theta}^*$:

    MTB> sort c5 c6

and then look through the worksheet to find the 97.5[th] value (of the 1000 bootstrap sample means). Report this value, and call it $\hat{\theta}^*_{.975}$.

(j) Repeat (i) to find the 2.5[th] percentile of the bootstrap sample means, calling this value $\hat{\theta}^*_{.025}$.

Notice that $P(\hat{\theta}^*_{.025} \leq \hat{\theta}^* \leq \hat{\theta}^*_{.975}) = .95$.

If we shift this distribution over by $\hat{\theta}$, we have

$$P(\hat{\theta}^*_{.025} - \hat{\theta} \leq \hat{\theta}^* - \hat{\theta} \leq \hat{\theta}^*_{.975} - \hat{\theta}) = .95,$$

which by the key result then gives:

$$P(\hat{\theta}^*_{.025} - \hat{\theta} \leq \hat{\theta} - \theta \leq \hat{\theta}^*_{.975} - \hat{\theta}) = .95.$$

Rearranging these terms gives:

$$P(2\hat{\theta} - \hat{\theta}^*_{.975} \leq \theta \leq 2\hat{\theta} - \hat{\theta}^*_{.025}) = .95.$$

And so the 95% bootstrap confidence interval for θ is $(2\hat{\theta} - \hat{\theta}^*_{.975}, 2\hat{\theta} - \hat{\theta}^*_{.025})$. As always, this confidence interval is specifying the plausible values of the parameter, θ, based on the observed sample estimate $\hat{\theta}$.

(k) Calculate this 95% bootstrap confidence interval for the population mean word length θ, based on the sample mean of C2 and your answers to (i) and (j).

(l) Follow the above analysis to determine a 90% bootstrap confidence interval for the population mean θ.

(m) For how many and what proportion of students in your class is the known population mean, 4.29, captured in the 90% bootstrap confidence interval? Is this close to what you expected? Explain.

Discussion: The 2.5th and 97.5th percentiles should have been approximately 3.5 and 6.2, leading to a 95% bootstrap confidence interval of approximately (3.4, 6.1), though of course your results will vary. This is reasonably close to the 95% t confidence interval, (3.26, 6.34). An advantage of the bootstrap confidence interval procedure is that it does not require any assumptions about the shape of the population distribution and can be used with a small sample. These results are also very similar to the "rough" bootstrap confidence interval found by estimating the standard error from the bootstrap samples and using the t critical value. In this example, since the sampling distribution was close to symmetric, we still obtain a roughly symmetric confidence interval. However, this percentile approach is more flexible and will lead to a non-symmetric interval if the bootstrap distribution is not symmetric.

The interpretation and properties of these bootstrap confidence intervals are the same as those from previous procedures. In particular, you should seen in (m) that a smaller confidence level produces a narrower interval. Your 90% bootstrap confidence interval should be approximately (3.7, 5.9).

Investigation 4-16: Comparison Shopping (cont.)

Recall the sample of grocery products obtained by the student project group (`shopping99.mtw`). Since the sample size is moderate, we might want to confirm that *t* confidence interval is valid. We can do so by comparing the resulting interval to a bootstrap confidence interval.

(a) Create a bootstrap distribution using the sample in C4 (you may need to remove the same unusual observations again). Remember to sample the same number of products as in this student sample, and to sample with replacement. Describe the resulting bootstrap distribution.

(b) Use this distribution to create a 95% percentile bootstrap confidence interval for the mean price difference in the population of items common to these two stores.

(c) How does the interval in (b) compare to the *t*-interval for the mean price difference?

Discussion: If the sample size is small and you don't feel comfortable applying the Central Limit Theorem, then this bootstrap procedure provides an alternative way to approximate the confidence interval for the parameter. The real beauty of this bootstrap procedure is that we can do it for any sample statistic.

Study Conclusions: While the sample showed a slight skewness, the sample size of 28 appears to be large enough for the *t* procedure to be valid. This is confirmed by the bootstrap percentile interval $(-.25, .01)$ being similar to the *t* confidence interval $(-.26, .02)$. Thus, we are 95% confident that the mean price difference for all products common to these two stores is at most 25 cents.

Investigation 4-17: Heroin Treatment Time

Hesketh and Everitt (2000) report on a study by Caplehorn and Bell (1991) that investigated the times that heroin addicts remained in a clinic for methadone maintenance treatment. The data in heroin.mtw includes the amount of time that the subjects stayed in the facility until treatment was terminated (C4). For about 37% of subjects, the study ended while they were still in the clinic (status = 0). Thus, their "survival time" has been "truncated." For this reason, we might not want to the estimate the mean survival time, but rather some other measure of a "typical" survival time. We will explore both the median and the *25% trimmed mean*. We will treat this group of 238 patients as representative of the population of heroin addicts.

(a) Produce and describe a histogram of the survival times for these patients.

(b) Use Minitab to simulate a bootstrap distribution of the median survival time (remember to sample 238 patients from C4, with replacement, but change the statistic from the sample mean to the sample median, storing the results in C8). Produce and describe a *dotplot* of the bootstrap distribution.

Discussion: This distribution is clearly not as "pretty" as what we are used to with sample means. It would not be well modeled by a normal or *t* distribution.

(c) Create and report a 95% percentile bootstrap confidence interval for the population median using the bootstrap distribution.

Discussion: Bootstrapping does not work as well with medians since there are not very many different possible values of the bootstrap median and you are relying on just one (or the average of 2) observation in the distribution. Another alternative is to focus on the "middle 50% of the distribution" by calculating a "25% trimmed mean." This statistic truncates the lower 25% of the distribution and the upper 25% of the distribution and then takes the mean of the remaining values (i.e., the middle 50% of the sample).

(d) Calculate the 25% trimmed mean of the treatment times in this sample:

```
MTB> sort c4 c7
MTB> copy c7 c8;
SUBC> include;        #to obtain the 60th and 178th values (the 25th and
SUBC> rows 60:178.    #75th percentiles for these 238 observations)
MTB> mean c8.
```

An advantage of using the trimmed mean rather than the sample mean is that it will be resistant to outliers. Unfortunately, we do not have a formula for the standard deviation of the trimmed mean. However, you should now believe that the standard deviation of the bootstrap distribution will provide a reasonable estimate and a percentile bootstrap confidence interval will allow us to estimate the population trimmed mean.

(e) Now create a macro to calculate the 25% trimmed mean for 1000 bootstrap samples. Create and describe a histogram of the resulting bootstrap distribution.

(f) Determine the 95% percentile confidence interval based on the bootstrap distribution. Write a one-sentence interpretation of the resulting interval.

(g) In fact, this sampling distribution should look quite symmetric so that you may use the t^* critical value instead, $t_{237} = 1.97$. Calculate a 95% confidence interval based on t^* and the standard deviation estimated by the bootstrap distribution.

Discussion: Since the treatment times are fairly skewed and most of the people still in the clinic have the longer treatment times, we might prefer a different statistic from the mean as a measure of a typical treatment time. Bootstrapping allows us to explore the sampling distribution of some of these other statistics, including the median and the 25% trimmed mean for which we do not have analytic results about the shape or spread of the sampling distribution. We do need to be cautious in using bootstrapping for the median, as it relies on a single value of the distribution, and statisticians recommend using it only for sample sizes of at least 100.

Study Conclusions: The distribution of treatment times was skewed to the right with a median of about 1 year (367.5 days) and interquartile range of 418.5 days. Because of the skewness in the data, the 25% trimmed mean was used as a measure of "average treatment time." Based on a bootstrap simulation, an approximate 95% confidence interval for the median treatment time in the population of heroin addicts is between 283 days and 420 days, while an approximate 95% confidence interval for the mean of the middle 50% of treatment times is 331 days to 419 days. These intervals include lower values than a *t*-interval for the population mean (368, 437), because the *t*-procedure is more strongly affected by the skewness in the data. We would like to generalize these data to all heroin patients, but would like more information to insure that this sample is representative.

Practice:

4-26) Golden Rectangles

The ancient Greeks made extensive use of the "golden rectangle" in art and architecture. They believed that a width-to-length ratio of .618 was aesthetically pleasing. Some have conjectured that American Indians used the same standard. Data were collected on the width-to-length ratios of beaded rectangles used by the Shoshoni Indians to decorate their leather goods. These data are stored in `shoshoni.mtw`.
(a) Produce and describe a histogram of these sample data.
(b) Comment on whether you believe the Central Limit Theorem will apply for these data.
(c) Produce a 95% bootstrap percentile confidence interval for the mean ratio.
(d) Produce a 95% bootstrap percentile confidence interval for the 25% trimmed mean.
(e) Write a paragraph summarizing these intervals. Based on these intervals, is there evidence that the preferred width-to-length ratio of the Shoshoni Indians differed from .618? Explain.

SUMMARY

In this chapter you focused primarily on the normal and t probability distributions as suitable models for different data sets and different sampling distributions. In particular, the Central Limit Theorem tells you that, under some technical conditions, the sampling distribution of a sample proportion and the sampling distribution of a sample mean are both well modeled by a normal distribution. This allowed you to derive test statistic and confidence interval formulas for making inferences about the population from which the sample is selected. It is important that you always check whether the technical conditions of a procedure are met before you apply it (and normal probability plots are a particularly useful tool here with quantitative data).

You saw a common structure to these procedures, the test statistic was always of the form:

$$\frac{sample\ estimate - hypothesized\ value}{standard\ error\ of\ estimate}$$

and most of the confidence interval procedures were always of the form:

$$sample\ estimate \pm crit.value \times (standard\ error\ of\ estimate).$$

The test statistic provides a standardized measure for how far, in terms of number of standard errors, that the sample result is from the hypothesized value. The larger (in absolute value) the test statistic, the smaller the p-value; the smaller the p-value, the stronger the evidence against the null hypothesis. We also formalized a structure to these tests of significance:

- defining the parameter
- stating the hypotheses
- checking the technical conditions of the procedure
- applying the procedure to determine the test statistic and p-value
- deciding to reject or fail to reject the null hypothesis
- stating a conclusion in context

Keep in mind that this structure and reasoning is the same, regardless of which procedure you choose. You were also reminded that there is always the chance we are making the wrong decision (Type I or Type II error) and that we should monitor these error probabilities.

With categorical data, we used the standard normal distribution to obtain the p-value and critical value, and with quantitative data, we used the *t* distribution. The table below summarizes the procedures and technical conditions discussed in this chapter. We hope you will use Minitab or applets to perform most of the calculations for you so that you may focus on interpreting the results and understanding the concepts of significance and confidence.

There are some calculations you will want to perform "by hand," such as calculating the probability of a Type II Error (or power against a specific alternative) and prediction intervals. It's important to distinguish between a confidence interval for a mean, and a prediction interval for an individual observation. You should also be aware of what factors affect the calculations, e.g., sample size.

When the Central Limit Theorem does not apply, there are newer, more computer-intensive methods such as bootstrapping that allow us to still make inferences. These are particularly useful for smaller samples or when we want to use a statistic other than the sample mean.

SUMMARY OF PROCEDURES TO MAKE INFERENCES ABOUT POPULATION PARAMETERS FROM A RANDOM SAMPLE:

	Binary (parameter π)	Quantitative (parameter μ)
Central Limit Theorem applies if	Data are SRS from population of interest and $n\pi \geq 10$ and $n(1-\pi) \geq 10$	Data are SRS from population of interest and either population distribution is normal or n is large
Central Limit Theorem says	Distribution of sample proportions will be approximately normal with mean π and standard deviation $SD(\hat{p}) = \sqrt{\pi(1-\pi)/n}$	Distribution of sample means will be (approximately) normal with mean μ and standard deviation $SD(\bar{x}) = \sigma/\sqrt{n}$.
Test statistic	$z_0 = \dfrac{\hat{p} - \pi_0}{\sqrt{\pi_0(1-\pi_0)/n}}$	$t_0 = \dfrac{\bar{x} - \mu_0}{s/\sqrt{n}}$
p-value	From standard normal distribution if $n\pi_0 \geq 10$ and $n(1-\pi_0) \geq 10$	From t_{n-1} distribution if $n \geq 30$ or sample approximately normal
Confidence interval	Wald: $\hat{p} \pm z^* \sqrt{\hat{p}(1-\hat{p})/n}$ if $n\hat{p} \geq 10$ and $n(1-\hat{p}) \geq 10$ adjusted Wald (95%): $\tilde{p} \pm 1.96\sqrt{\tilde{p}(1-\tilde{p})/(n+4)}$ if $n \geq 5$.	CI for μ: $\bar{x} \pm t^*_{n-1} s/\sqrt{n}$ if $n \geq 30$ or sample approximately normal (robust) prediction interval: $\bar{x} \pm t^*_{n-1} s\sqrt{1+1/n}$ if $n \geq 30$ or sample approximately normal (not robust)
If Central Limit Theorem does not apply consider	Binomial distribution	Bootstrapping

TECHNOLOGY SUMMARY

- In Minitab, you learned how to
  - Overlay a distributional model on a histogram
  - Create probability plots to assess the fit of a model to data
  - Calculate probabilities from a normal distribution or t distribution, including the standard normal distribution, and how to calculate inverse cumulative probabilities
  - Perform one-sample z-procedures for inference about a population proportion
  - Perform one-sample t-procedures for inference about a population mean
  - Simulate sampling distributions and bootstrap distributions
- You learned how to use the normal probability calculator applet and the test of significance calculator.
- You also used technology to explore concepts such as sampling distributions and confidence. We hope you will maintain these visual pictures in your memory as these applets (Reeses' Pieces, Simulating Confidence Intervals) were for illustrative purposes, not for generic analysis.

CHAPTER 5: COMPARING TWO POPULATIONS

In chapter 3 you learned about inference procedures for making conclusions about a population parameter or probability when you have randomly selected a sample from that population or process. In chapter 4, you learned about the utility of the normal distribution and the *t* distribution in modeling the sampling distribution of a statistic, especially when the sample size is reasonably large. In this chapter you will essentially apply these same ideas to the goal of comparing two populations. You will first focus on comparing two different population proportions, where you have randomly selected an independent sample from each population, and then you will focus on comparing two population means, as well as other sample statistics. As you have done before, you will begin by simulating the corresponding sampling distribution and then seeing whether the probability models from the last chapter provide reasonable approximations of these sampling distributions. While we are comparing two groups as we did in chapters 1 and 2, here the randomness arises from drawing the samples from the population, not in the assignment of subjects to groups. Still, you will see that the methods developed here also help us answer the questions we examined in the first two chapters—comparing treatment groups in a randomized experiment. Only in the latter case will we be able to draw any cause and effect conclusions. Only in the former case will we be able to generalize the differences in the samples to differences in the populations. When a study does both—randomly selects a sample from a population and randomly assigns subjects to groups—then we can both draw a causal conclusion and generalize the results to the larger population.

Section 5-1: Comparing two samples on a categorical response
> Investigation 5-1: Newspaper Credibility Decline—two proportion sampling distribution
> Investigation 5-2: Newspaper Credibility Decline (cont.)—two-sample *z*-test and *z*-intervals
> Investigation 5-3: Sleepless Drivers—CI for odds ratio

Section 5-2: Randomized experiments revisited
> Investigation 5-4: Letrozole and Breast Cancer—normal approximation for Fisher's Test

Section 5-3: Comparing two samples on a quantitative response
> Investigation 5-5: NBA Salaries—two-sample *t* procedures (Minitab Exploration)
> Probability Detour
> Investigation 5-6: Handedness and Life Expectancy—effect of sample size and SD
> Investigation 5-7: Comparison Shopping (cont.)—two- vs. one-sample with paired data

Section 5-4: Randomized experiments revisited
> Investigation 5-8: Sleep Deprivation (cont.)— *t* approximation to randomization distribution

Section 5-5: Other statistics
> Investigation 5-9: Heart transplants and survival—bootstrap difference in medians and randomization test for medians

SECTION 5-1: COMPARING TWO SAMPLES ON A CATEGORICAL RESPONSE

In this section you will focus on comparing two populations with respect to a categorical response variable. In comparing the samples numerically and graphically, you will be able to rely on the procedures learned in chapter 1: segmented bar graphs, differences in proportions, and odds ratios. Since we will have random samples from the populations of interest, you will then explore techniques that allow us to generalize from these samples to the larger populations, something we couldn't always do in chapter 1.

Investigation 5-1: Newspaper Credibility Decline

With the proliferation of the Internet and 24-hour cable news outlets, it has become much easier for people to hear much more information, much more quickly. However, this has led to speculation that news organizations attempt to convey information before it has been properly verified in an effort to feed our impatience. *USA Today* reported that newspapers appear to be losing credibility over time (March, 2004). They cited a nationwide sample of 1,002 adults, 18 years or older, interviewed via telephone (under the direction of Princeton Survey Research Associates) during the period May 6-16, 2002. One of the questions asked was:

> Please rate how much you think you can BELIEVE each organization on a scale of 4 to 1. On this four point scale, "4" means you can believe all or most of what the organization says. "1" means you believe almost nothing of what they say. How would you rate the believability of (READ ITEM. ROTATE LIST) on this scale of 4 to 1?

The interviewer then asked this question for several different news organizations (e.g., *USA Today*, NPR, MSNBC).

(a) Why is it important for the interviewer to rotate the list of organizations?

In this investigation, we will focus on people's responses to the credibility issue for "The daily newspaper you are most familiar with." About 6-7% of respondents said they were not able to rate their daily newspaper. Of the 932 respondents who were able to rate the daily newspaper they were most familiar with, 587 rated the paper as "largely believable" (a 3 or a 4 on the scale). When the same question was asked four years earlier (May 7-13, 1998), 922 said they could rate their daily paper and of those, 618 rated the paper as "largely believable."

(b) Identify the observational units and the variables of interest. Which variable would you consider the explanatory variable and which the response variable? Is this an observational study or an experiment? Define the samples and the populations.

    Observational units:

    Variables:

Explanatory variable:
Response variable:

Observational study or experiment:

Samples:

Populations:

(c) Organize these data into a two-way table. Produce and comment on a segmented bar graph to compare the sample proportions rating their daily newspaper as largely believable in these two years. Is there evidence in these samples that the believability has decreased?

(d) Is it plausible that we would see the sample proportion that rate their daily newspaper as largely believable in 2002 drop by .04 (the difference observed in this study), even if the overall population proportion that feels this way has not changed from 1998? Explain.

(e) Let X represent the number of respondents in a randomly selected sample of 922 people who rate their daily paper as largely believable in 1998. Is X (approximately) a binomial random variable? Justify your conclusion.

(f) Let Y represent the number of respondents in a randomly selected sample of 932 people who rate their daily paper as largely believable in 2002. Is Y (approximately) a binomial random variable?

(g) Let $Z = X-Y$. Is Z (approximately) a binomial random variable? Why or why not?

(h) Define $\pi_1-\pi_2$ to be the difference in the proportion who would rate their daily paper as largely believable between these two years. If there were no difference in people's rating of the believability between these two years, what would this imply about $\pi_1 - \pi_2$? What if the believability has decreased?

(i) Turn your answers in (h) into a null hypothesis statement and an alternative hypothesis statement.

Even though X and Y both have binomial distributions (approximately, because the population is finite but huge), their difference X-Y does not follow a binomial distribution. Therefore, we will use simulation to investigate the behavior of the random variable
$$W = X/n_1 - Y/n_2 = \hat{p}_1 - \hat{p}_2,$$
assuming that the null hypothesis of no decline is true. For this simulation, suppose the population proportions are equal to some common value, say $\pi_1 = \pi_2 = .65$. Call this common value π.

(j) Use Minitab to generate 1000 observations from a binomial distribution with sample size $n_1 = 922$ and success probability $\pi = .65$, storing the results in C1 and sample proportions in C2:

```
MTB> random 1000 c1;
SUBC> binom 922 .65.
MTB> let c2=c1/922
MTB> name c2 'phat98'
```

Then generate 1000 observations from a binomial distribution with $n_2 = 932$ and $\pi = .65$, storing the results in C3 and then computing the resulting sample proportions and storing those results in C4 and naming the column phat02.

Produce and describe numerical and graphical summaries for each column.

	X_1	$\hat{p}_1$	X_2	$\hat{p}_2$
Mean				
Std Dev				

Descriptions:

Are the mean and standard deviation of each set of observations close to what you would predict? Explain.

(k) Would it be reasonable to model the distribution of $\hat{p}_1$ with a normal distribution? With what mean and standard deviation? What about $\hat{p}_2$?

(l) Now store the differences between the columns of sample proportions in C5. Produce and describe numerical and graphical summaries of these differences. Record the mean and standard deviation. Use a normal probability plot to assess whether these differences appear to be reasonably modeled by a normal distribution. Comment on what you observe.

Mean: Standard deviation:

Description:

Normal model?

(m) Recall that the proportion of people who rated their daily newspaper as largely believable was .04 higher in 1998 than in 2002. What percentage of your 1000 simulated differences is larger than the actual difference observed between these two polls? What conclusion would you draw from this empirical p-value?

Study Conclusion: The sample from 1998 shows a slightly higher percentage of people rating their daily newspaper as largely believable (difference in sample proportions .04). Since these are large, independent samples, we can apply a two-sample z-procedure to see if this difference is statistically significant. There is moderate evidence that a difference this large would not have occurred by the random sampling process alone. The simulation above assumes no difference between the two populations ($\pi = .65$ for both years) and shows that it is somewhat surprising (one-sided empirical p-value < .05) to have a difference in proportions of at least .04 between 1998 and 2002. However, we need a procedure that doesn't assume a numerical value for this common proportion.

Discussion: This simulation not only reveals that the observed difference in sample proportions is statistically significant, it also suggests that the normal probability model serves a useful approximation for the sampling distribution of the statistic $\hat{p}_1 - \hat{p}_2$. Since the difference between two binomial random variables is not itself binomial, you will now explore an alternative probability model for these differences. This normal model can be used to calculate a test statistic and approximate p-value and also to calculate a confidence interval for the difference in the population proportions.

Investigation 5-2: Newspaper Credibility Decline (cont.)

(a) Reconsider (or reproduce) your simulation results from the previous investigation. How do the mean and standard deviation of the differences in C5 compare to the means and standard deviations you suggested in Investigation 5-1(k)?

Probability Rule: If X has a normal distribution with mean μ_x and standard deviation σ_X and Y has a normal distribution with mean μ_Y and standard deviation σ_Y, then the difference X-Y also has a normal distribution. The mean of X-Y is $\mu_X - \mu_Y$, and the standard deviation of X-Y is $\sqrt{\sigma_X^2 + \sigma_Y^2}$ *provided that* X and Y are independent.

Because our sample sizes are large, the binomial distributions of both X and Y can be approximated by a normal distribution. Thus, the distribution of their difference also approximately follows a normal distribution. A key observation is that the variance of the differences is the *sum* of the variances, as long as the random variables can be considered *independent*. In this case, we are considering the data a random sample of respondents in 1998 and an independent random sample of respondents in 2002. This assertion would be problematic if the observations had been *paired* in some way, e.g., asking the same people in both years.

(b) Now define the random variable $\hat{p}_1 - \hat{p}_2 = X/n_1 - Y/n_2$, where X is binomial$(n_1, \pi_1)$ and Y is binomial(n_2, π_2). Use probability rules to derive the generic expressions for the theoretical expected value and variance of this random variable, under the null hypothesis assumption that $\pi_1 = \pi_2$. [*Hint*: Remember to work with variances, not standard deviations, and look back to the expected value and variance formulas for a binomial distribution from chapter 3.]

(c) Verify that the mean and standard deviation of your 1000 simulated differences are close to the theoretical values, again assuming that $\pi_1 = \pi_2 = .65$ as you did in the simulation.

In working with the difference between two binomial random variables, we will approximate the sampling distribution of $\hat{p}_1 - \hat{p}_2$ with a normal distribution having mean $\pi_1 - \pi_2$ and standard deviation $\sqrt{\dfrac{\pi_1(1-\pi_1)}{n_1} + \dfrac{\pi_2(1-\pi_2)}{n_2}}$. Under the null hypothesis H_0: $\pi_1 - \pi_2 = 0$, we let $\pi = \pi_1 = \pi_2$, and the standard deviation simplifies to $\sqrt{\pi(1-\pi)\left(\dfrac{1}{n_1} + \dfrac{1}{n_2}\right)}$. We will consider the normal model appropriate if it is appropriate for each binomial random variable, namely $n_1\pi_1 \geq 5$, $n_1(1-\pi_1) \geq 5$ and $n_2\pi_2 \geq 5$, $n_2(1-\pi_2) \geq 5$.

However, to perform these calculations we would need to know π_1, π_2, or π. So again we estimate the standard deviation of $\hat{p}_1 - \hat{p}_2$ using the sample data. When the null hypothesis is true, we are assuming the samples come from the same population, so we pool the two samples together to estimate the common population proportion of successes. That is, we estimate π by looking at the ratio of the total number of successes to the total sample size:

$$\hat{p} = \frac{X+Y}{n_1+n_2} = \frac{n_1\hat{p}_1 + n_2\hat{p}_2}{n_1+n_2}$$

Then use we use this value to calculate the standard error of $\hat{p}_1 - \hat{p}_2$ to be:

$$SE(\hat{p}_1 - \hat{p}_2) = \sqrt{\hat{p}(1-\hat{p})\left(\frac{1}{n_1} + \frac{1}{n_2}\right)}$$

(d) Use these results to suggest the general formula of a test statistic and a method for calculating a p-value to test H_0: $\pi_1 - \pi_2 = 0$ (also expressed as H_0: $\pi_1 = \pi_2$) versus the alternative H_a: $\pi_1 - \pi_2 > 0$.

(e) Carry out the test specified in (d) using the result that 618 of the 922 respondents in 1998 and 587 of the 932 respondents in 2002 said their daily paper was largely believable.

$\hat{p}$:

$SE(\hat{p}_1 - \hat{p}_2)$:

test statistic:

p-value:

Is the standard error close the empirical standard deviation from your simulation results? Is the p-value close to the empirical p-value from your simulation results?

(f) Do the sample data provide convincing evidence that the public is losing trust in their daily paper?

Discussion: You have now learned two ways to conduct an approximate test for comparing proportions between two samples, one based on simulation and the other based on a normal approximation. The exact distribution of the difference between two binomial random variables is not convenient to work with, but Fisher's Exact Test can be used when the sample sizes are small. However, determination of confidence intervals and rejection regions are not as straight forward with the hypergeometric distribution, so we will apply this normal model when the technical conditions are met.

As always, a test of significance only indicates whether a difference is unlikely to have occurred by chance. If this test turns out to be significant, it again makes sense to use a confidence interval to estimate the magnitude of the difference between the two groups.

To estimate the size of the difference in the population proportions we can follow this test with a confidence interval for $\pi_1 - \pi_2$. We are no longer assuming these two population proportions are equal, so we will calculate the standard error without pooling the samples:

$$SE(\hat{p}_1 - \hat{p}_2) = \sqrt{\frac{\hat{p}_1(1-\hat{p}_1)}{n_1} + \frac{\hat{p}_2(1-\hat{p}_2)}{n_2}}$$

(g) Calculate this standard error from the data for this study. Then use this standard error to construct an approximate 90% confidence interval for the difference in the population proportions who rate their daily paper as largely believable between these two years. [*Hint*: Use the normal distribution to calculate the critical value.]

Standard error:

90% confidence interval:

(h) Interpret the interval you calculated in (g).

(i) Explain what is meant by "90% confidence" in (g).

(j) Is the interval you calculated in (g) consistent with your conclusion from Investigation 5-1, part (m) and Investigation 5-2, part (f)? Explain.

(k) Calculate a 95% confidence interval for $\pi_1-\pi_2$. How does it compare to the 90% interval?

We can also calculate a confidence interval using a Wilson adjustment for the sample proportions as we did in chapter 4. This time put *one* additional success and one additional failure in each sample: $\tilde{p}_1 = (X_1+1)/(n_1+2)$ and $\tilde{p}_2=(X_2+1)/(n_2+2)$. The point estimate will be $\tilde{p}_1-\tilde{p}_2$ and we will use these values in the SE calculation as well.

$$ SE(\tilde{p}_1-\tilde{p}_2) = \sqrt{\frac{\tilde{p}_1(1-\tilde{p}_1)}{n_1+2} + \frac{\tilde{p}_2(1-\tilde{p}_2)}{n_2+2}} $$

An approximate 95% confidence interval for $\pi_1-\pi_2$ will then be $\tilde{p}_1-\tilde{p}_2 \pm 1.96\ SE(\tilde{p}_1-\tilde{p}_2)$
(other confidence levels are possible but the adjustments are not presented here)
This adjustment will be most useful when the proportions are close to 0 or 1, i.e., when the conditions for the conventional two-sample z procedure are not met.

(l) Calculate the confidence interval using the Wilson adjustment. How does it compare to your confidence interval in (g)?

Study Conclusions: We have moderately strong evidence that the population proportion that would rate their daily newspaper as largely believable in 1998 was larger than the population proportion in 2002 (one-sided p-value = .036). However, this difference is fairly small. The 90% confidence interval for the difference in the population proportions suggests a 0.4% to 7.7% decrease between 1998 and 2002. But the 95% confidence interval suggests that zero is a plausible value for $\pi_1-\pi_2$. (The adjusted interval is very similar since the sample sizes are large and the sample proportions are near .5). We feel comfortable generalizing this result back to the populations assuming the nationwide sample gathered by this professional polling organization is representative. This was an observational study so we cannot identify the reason behind the decrease in credibility from these sample data.

Brief summary of two-sample *z*-procedures

If we have independent random samples and at least 5 successes and 5 failures in each group, then the test statistic $z_0 = (\hat{p}_1 - \hat{p}_2)/\sqrt{\hat{p}(1-\hat{p})\left(\frac{1}{n_1} + \frac{1}{n_2}\right)}$ can be compared to the standard

normal distribution and $(\hat{p}_1 - \hat{p}_2) \pm z^* \sqrt{\frac{\hat{p}_1(1-\hat{p}_1)}{n_1} + \frac{\hat{p}_2(1-\hat{p}_2)}{n_2}}$ is an approximate C% confidence

interval for $\pi_1 - \pi_2$ (where $-z^*$ is the $(100-C)/2^{\text{th}}$ percentile from the standard normal distribution). The coverage properties of this interval can be slightly improved by using a Wilson adjustment.

Minitab

Choose Stat > Basic Statistics > 2 Proportions.

- Click Summarized data
- Enter the sample size (trials) and number of successes for Group 1. Note: You will have to convert the *proportion* of successes to the *number* of successes (events) first.
- Enter the sample size and number of success for Group 2
- Click the Options button
- Select the box to "Use pooled estimate of p for test"
- Specify the alternative to be "greater than"
- Click OK twice.

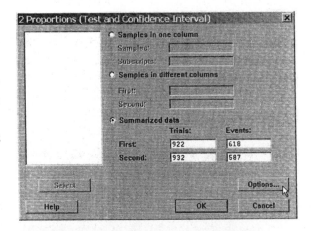

- To obtain the confidence interval, choose Stat > Basic Statistics > 2 Proportions again and change the direction of the alternative back to "not equal to" under the Options button.

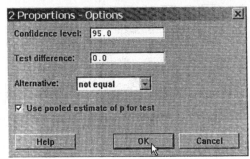

- To use Minitab to perform the Wilson adjustment, enter the adjusted counts (increase each by 1) and the adjusted sample sizes (increase each by 2).

Practice:

5-1) The Governator
In 2003, the state of California conducted their first recall election. Over one hundred candidates vied to replace Gray Davis as Governor, with actor Arnold Schwarzenegger winning 49% of the votes cast. In the days before the election, newspapers had published accusations of inappropriate treatment of women by Schwarzenegger, so some suspected that his support from women would be weaker than his support from men. Before the polls close, pundits try to predict the election results by using "exit polls," asking people as they leave the voting booth who they voted for. In one such poll, CNN interviewed 2023 men and 2191 women. Forty-nine percent of the men and 43% of the women said they voted for Arnold.
(a) Specify the populations and parameters of interest.
(b) Produce and comment on a segmented bar graph to compare the proportion voting for Arnold for men and women. Is there evidence in these samples that men were more likely than women to say they voted for Arnold?
(c) Define $\pi_m - \pi_f$ to be the difference in the proportion who would say they voted for Arnold in the population of male California voters and in the population of female California voters. If there were no difference between men and women in whether they would say they voted for Arnold, what would this imply about $\pi_m - \pi_f$? What if men are more likely to say they voted for Arnold? Turn your answers into a null hypothesis statement and an alternative hypothesis statement.
(d) Are the conditions met for us to apply the methods of this section? Explain.
(e) Calculate the test statistic and approximate p-value. Write a paragraph summarizing your conclusions.
(f) Calculate and interpret a 90% confidence interval for the difference in the population proportion of men and women who voted for Arnold.

5-2) Sample Size Effects
(a) Would the sample proportion of 49% have been significantly larger than 43% (statistically) if the sample sizes had been 100 men and 100 women? Report the z-test statistic and (one-sided) p-value. Also indicate whether the difference is significant at the .10 level, at the .05 level, and at the .01 level.
(b) Repeat (a) if the sample sizes had been 600 men and 600 women (assuming the same sample proportions).
(c) What happens to the p-value of the test as the sample size increases? Explain why this makes sense.
(d) Determine the smallest sample size (assuming the same number in each group) for which the difference between 43% and 49% is significant at the .05 level, using a one-sided test.
(e) How would the confidence interval from Practice 5-1(f) differ if the samples sizes had been 100 men and 100 women (assuming the same sample proportions)? Explain.
(f) If we assume about 50% of each gender voted for Arnold, what sample size (assuming the same sample size in each group) is necessary for the margin of error (with 90% confidence) to be at most 5%?

5-3) Penny Thoughts

Students in a statistics class were asked whether they would vote to retain or abolish the penny. Of 10 men, 6 voted to retain the penny. Of 15 women, 14 voted to retain the penny. Treat these as independent random samples of male statistics students and female statistics students at this university.

(a) Calculate an approximate 95% confidence interval (without the Wilson adjustment) for the difference in the proportion of male students at this university who would vote to retain the penny and the proportion of female students at this university who would vote to retain the penny.

(b) Are the technical conditions for this procedure satisfied? Explain.

(c) Repeat (a) using the Wilson adjustment. How do the intervals compare?

(d) Why should you be cautious about interpreting this interval for the population of all students at this university?

Investigation 5-3: Sleepless Drivers

Connor et. al. (British Medical Journal, May 2002) reported on a study that investigated whether sleeplessness is related to car crashes. The researchers identified all drivers or passengers of eligible light vehicles who were admitted to a hospital or died as a result of a car crash on public roads in the Auckland, New Zealand region between April 1998 and July 1999. Though cluster sampling, they identified a sample of 571 drivers who had been involved in a crash resulting in injury, and a sample of 588 drivers who had not been involved in such a crash was identified. These individuals were selected through cluster sampling to be representative of people driving on the region's roads during the study period. The researchers asked the individuals if they had a full night's sleep any night during the previous week.

(a) Identify the observational units and variables in this study. Which variable would you consider the explanatory variable and which the response variable?

(b) Is this an observational study or an experiment? Explain.

(c) Would this be considered a case-control, a cohort, or a cross-classified design? Explain.

(d) Is it reasonable to consider these as two independent random samples? If so, from what populations? Explain.

(e) Suppose we define the parameter π_1 to be the proportion of drivers who had not experienced a full night's sleep in the previous week that had a car accident and π_2 to be the proportion of drivers who had experienced a full night's sleep that had an accident. Would it be appropriate to estimate π_1-π_2 from these data? Explain. [*Hint*: Consider some of the study design issues discussed in Section 1-2.]

We cannot estimate this parameter from the data because the distribution of the accident variable was controlled by the researchers. Therefore, we will instead consider the population odds ratio as the parameter of interest. Define τ to be the population odds ratio of a car accident for the sleepless group compared to the "full night sleep" group.

(f) State the null and alternative hypotheses for testing whether this odds ratio is greater than 1. State in words what these hypotheses imply about the association between sleeplessness and occurrence of car accidents.

The researchers found that 61 of the 571 "case" drivers and 44 of the 588 "control" drivers had not gotten a full night's sleep in the previous week.

(g) Organize these sample data into a two-way table:

	No full night's sleep in past week	At least one full night's sleep in past week	Sample sizes
"case" drivers (crash)			571
"control" drivers (no crash)			588

(h) Produce and discuss numerical and graphical summaries of these sample data, including the sample odds ratio, denoted by $\hat{\tau}$. What do these summaries reveal? Does the sample odds ratio appear to be extreme?

In order to evaluate whether this sample odds ratio is extreme, we need information about the sampling distribution of the odds ratio when the population proportions are the same (and so the population odds ratio $\tau = 1$). You will use simulation to approximate this sampling distribution when this null hypothesis is true, and then assess whether 1.48 is a larger sample odds ratio than would typically occur by chance. The simulation below assumes that 9% (the pooled estimate from the sample data) of drivers in the population did not get a full night's sleep in the previous week (for both the cases and the controls). From this population, we will sample 571 "case" drivers and 588 "control" drivers to mimic the researchers' study, under the hypothesis of no association between the two variables.

(i) Use Minitab to randomly generate 1000 observations from a binomial distribution with $\pi = .09$ and $n = 571$ (Calc > Random Data > Binomial), storing the results in C1, and 1000 observations from a binomial distribution with $\pi = .09$ and $n = 588$, storing the results in C2. Then calculate the odds ratio for each row (pair of samples) as follows:

```
MTB> let c3=(c1*(588-c2))/(c2*(571-c1))
```

Produce and discuss numerical and graphical summaries for these simulated odds ratio values. Are they reasonably modeled by a normal distribution? [*Hint*: Examine a normal probability plot.] Is the mean close to what you would have predicted? Explain.

Mean: Standard deviation:

Description:

Normal?:

Mean close to prediction?

(j) How often did the simulation produce an odds ratio at least as extreme as the 1.48 value observed by the researchers? What conclusion do you draw from this empirical p-value?

Study Conclusions: The sample odds ratio of 1.48 indicates that the odds of a sleepless driving having a crash were about 50% higher than those for a well-rested driver in this sample. The empirical p-value (less than 5%) provides moderately strong evidence that such an extreme value for the sample odds ratio is unlikely to have arisen by chance alone if the proportion of drivers with sleepless nights was .09 for both the population of "cases" and the population of "controls."

Discussion: It would be nice to follow this simulation up with a probability model for the distribution of the sample odds ratio that did not require us to assume a value for the population proportion as we did above. However, the distribution of the simulated odds ratios in C3 is clearly skewed to the right and so will not be well modeled by a normal distribution. This makes sense because the range of possible values for the sample odds ratio is not symmetric around 1 (since it is bounded by zero on the left and unbounded on the right). Thus, to be able to use a probability model, we either need to determine an appropriate probability model for these odds ratios or we can transform the odds ratio into another statistic that does follow a common probability distribution. Next you will study the latter strategy.

(k) Use Minitab to determine the *log odds ratio* for your 1000 simulated samples:

```
MTB> let c4=log(c3)
```

[*Note*: Minitab assumes the natural log unless you type "logten(c3)", but either base will suffice here.] Produce and discuss numerical and graphical summaries for these log odds ratios. Are the log odds reasonably modeled by a normal distribution? Is the mean close to what you would have predicted? Explain.

Mean: Standard deviation:

Description:

Normal?:

Mean close to prediction?

Although the sampling distribution of the sample odds ratio is not normal, the sample log-odds values are approximately normally distributed. Thus, we can conduct a test and construct a confidence interval for the population log-odds ratio using the normal distribution. The standard error of the sample log-odds ratio (using the natural log) is given by the expression:

$$SE(\text{log-odds}) = \sqrt{\frac{1}{a} + \frac{1}{b} + \frac{1}{c} + \frac{1}{d}}$$ where a, b, c, and d are the four table entries.

(l) Calculate the log odds ratio for the sample data in this driver sleepiness study, and then calculate the standard error, $SE(\text{log-odds})$. Verify that the standard deviation of your empirical sampling distribution is close to this value.

(m) Construct a 90% confidence interval for the population log-odds based on the sample data. [*Hint*: First calculate the sample value of the log-odds. Then go 1.645 standard errors on either side of that value.]

(n) Exponentiate these two endpoints of the interval to get a 90% confidence interval for the population odds ratio. Does your interval contain the value one? Discuss the implications of the interval containing one or not.

Study Conclusions:

The proportions of drivers who had not gotten a full night's sleep in the previous week were .107 for the case group of drivers who had been involved in a crash, compared to .075 for the control group who had not. Because these proportions are small, and because of the awkward roles of the explanatory and response variables in this study (we would much rather make a statement about the proportion of sleepless drivers who are involved in crashes), the odds ratio is a more meaningful statistic to calculate. The sample odds of having missed out on a full night's sleep were 1.48 times higher for the case group than for the control group. By the invariance of the odds ratio, we can also state that the sample odds of having an accident are 1.48 times higher for those who do not get a full night sleep than those who do. Your initial simulation results found moderately strong evidence (one-sided empirical p-value $\approx$ .03) that the population of drivers involved in crashes is more likely to have gone without a full night's sleep. A 90% confidence interval for the population odds ratio extends from 1.05 to 2.08. This interval provides statistically significant evidence that the population odds ratio exceeds one and that the odds of having an accident are about 1 to 2 times higher for the sleepless drivers than for well-rested drivers. We cannot attribute this association to a cause-and-effect relationship because this was an observational (case-control) study.

A confidence interval for a population odds ratio τ is:

$$\exp\left(\log(\hat{\tau}) \pm z^* \sqrt{\frac{1}{a} + \frac{1}{b} + \frac{1}{c} + \frac{1}{d}}\right), \text{ where } \hat{\tau} \text{ denotes the sample odds ratio.}$$

Practice:

5-4) Sleepless Drivers (cont.)
The New Zealand researchers also recorded whether a driver had obtained less than 5 hours of sleep in the previous 24 hours. They found that 65 of 529 drivers in the case group had less than 5 hours, compared to 30 of 584 drivers in the control group.
(a) Calculate a 90% confidence interval for the population odds ratio, τ. Remember to provide an interpretation of the sample odds ratio and of this confidence interval.
(b) Does this interval provide convincing evidence that the odds of a crash are higher for the group with less sleep? Explain.

5-5) Women's College Basketball
A February, 2004 Harris Poll asked a nationwide sample of 2,204 adults whether or not they followed college basketball. When asked about women's basketball, 7% said they did follow. When the same question was asked in December 1998, 8% of 1,005 respondents said they did follow women's college basketball.
(a) Is there evidence of a change in the women's basketball following over this 6-year period? (Examine both the difference in proportions and the odds ratio.)
(b) Construct and interpret a 95% confidence interval for the population odds ratio, τ.

SECTION 5-2: RANDOMIZED EXPERIMENTS REVISITED

The above activities focused on comparing two sample proportions where the samples were independently selected from two populations. In chapter 1 you also compared two "sample proportions" when you examined the proportion of successes between two treatment groups. In this investigation you will explore the similarity of the analyses in these two cases and the differences in the scope of conclusions that should be drawn.

Investigation 5-4: Letrozole and Breast Cancer

The November 6, 2003 issue of the *New England Journal of Medicine* reported on a study of the effectiveness of letrozole in postmenopausal women with breast cancer who had completed five years of tamoxifen therapy. The researchers wanted to know if letrozole increased the rate of disease-free survival. Over 5000 women were enrolled in the study. The women were randomly assigned to receive either letrozole or a placebo. The primary end result studied was disease-free survival. The article reported that 92.8% of the 2575 women who received letrozole achieved disease-free survival, compared to 86.8% of the 2582 women in the placebo group.

(a) Do you suspect that these 5157 women were randomly selected from all women with breast cancer who had completed five years of tamoxifen therapy? Do you suspect the sample of women who received letrozole is a random sample from some larger population? Explain.

(b) Is this an experiment or an observational study? Explain.

Discussion: As often happens with experiments, we are interested in seeing if the observed difference in the treatment groups is larger than we would expect from the randomization process alone, but we may not be willing to generalize to a larger population. Thus, instead of defining population parameters, we will define the parameter δ to represent the underlying "treatment effect." If there is no treatment effect, then δ equals zero, but we might also conjecture that δ is positive, negative, or different from zero.

(c) State the null and alternative hypotheses for this study in terms of δ and in words.

(d) Explain the consequences of a Type I error and of a Type II error in this study.

(e) Can you use Fisher's Exact Test to assess the significance of the increase in disease-free survival for the letrozole treatment?

(f) Below is the two-way table of these results. Based on this table, calculate the odds ratio for disease-free survival of letrozole users compared to the placebo group, and then carry out Fisher's Exact Test, using the hypergeometric distribution in Minitab to calculate the p-value assuming there is no effect from the drug. What conclusion can you draw about the effectiveness of letrozole?

	Letrozole	Placebo	Total
Disease-free survival	2390	2241 (k2)	4631
Not disease-free survival	185	341 (k3)	526
Total	2575	2582	5157

While you can use Minitab to carry out this calculation with ease, you can imagine the difficulty of performing such a calculation "by hand." When the sample size is so large, we can consider alternative calculations that, before computers, were much less computationally intensive. The following steps will help you explore suitable probability models to describe the randomization distribution of the log odds ratio for this study.

(g) Open the Minitab worksheet `letrozole.mtw`. Column 1 contains 526 zeros (representing the patients who did not have disease-free survival) and 4631 ones (representing patients with disease-free survival). Column 2 contains 2575 zeros (the letrozole group) and 2582 ones (the placebo groups). The following Minitab commands simulate the randomization of these subjects to the two treatment groups, storing the difference in sample proportions and the log-odds. You should create a Minitab macro with these commands:

```
sample 5157 c2 c5          #reassigns the treatment groups
let c6=c1*c5               #equals one if placebo and successes
let k2=sum(c6)             #counts number of successes in placebo group
let c7(k1) = k2/2582 -(4631-k2)/2575   #difference in proportions
let k3=2582-k2             #number of failures in placebo group
let c8(k1) = log(k2*(526-k3)/(k3*(4631-k2)))    #log odds ratio
let k1=k1+1
```

Initialize `k1` and execute the macro 1000 times. Produce and describe graphical and numerical summaries of the empirical randomization distribution of the difference in proportions and of the log odds ratio. Do they appear to be well modeled by a normal distribution?

(h) The observed sample odds ratio in the letrozole study is $\hat{\tau}$ =1.966. Calculate the empirical p-value by seeing how many of the simulated log odds ratios are greater than or equal to *log*(1.966), the observed log-odds ratio in the letrozole study. [*Hint*: Use `let c9=(c8>=log(1.966))` and then `tally c9`.]

When the sample sizes are large, the sampling distribution of the difference in the conditional proportions and the sampling distribution for the log-odds for a randomized comparative experiment can each be modeled by a normal distribution. We will consider the sample sizes large enough when there are at least 5 observations in each cell of the two-way table.

(i) Use Minitab to carry out a two-sample z test for these data. How does the p-value compare to the one from Fisher's Exact Test that you found in (f)? [*Minitab hint*: Under Stat > Basic Statistics > 2 Proportions, you can now choose the first option, samples in one column, and enter C1 as the samples and C2 as the subscripts.]

(j) Calculate (by hand) and interpret a 99% confidence interval for the odds ratio of survival with letrozole versus placebo.

Study Conclusions: As you first saw in chapter 1, there is very strong evidence that the proportion of disease-free survivors is larger for those taking letrozole than those taking placebo ($\delta > 0$). The p-value is virtually zero, and now we also see that the test statistic is 7.18, revealing that the sample results are a whopping 7.18 standard deviations away from expected. We are 99% confident that the odds ratio is between 1.54 and 2.52, indicating that the odds of disease-free survival roughly double when taking letrozole. We can attribute this change to the letrozole since this was a randomized, double-blind, comparative experiment. Technically, the subjects were not randomly selected from a population so we should consider whether they are likely to be representative of all women with breast cancer before generalizing this result beyond those women in the study. If you do not have reason to believe that this sample was randomly selected from the population of women with breast cancer, then you must limit your conclusions to the differences in these treatment groups. Many pharmaceutical companies have learned this lesson the hard way when a drug found to be effective in Caucasian males worked differently in women, elderly, and children. Information about the representativeness of these subjects can be gained from examining demographic and other background variables recorded at the start of the study.

Discussion: While the initial discussion in this chapter focused on random samples from independent populations, this investigation has again returned to the process of randomly assigning subjects to treatment groups as you saw in chapter 1. The question of interest was whether the differences observed in the response across the explanatory variable groups could have plausibly arisen due to the randomization process alone (H_o: $\delta = 0$, no treatment effect). In chapter 1 you used Fisher's Exact Test with the hypergeometric distribution to calculate the p-value exactly. In this investigation you saw that when the sample sizes are large, the randomization distribution is well approximated by a normal distribution. Historically, this allowed analysts to approximate the p-value when hypergeometric calculations would have been cumbersome, as well as to calculate test statistics and confidence intervals.

Summary of two sample procedures for a categorical response

To test H_0: $\pi_1 - \pi_2 = 0$ or H_0: $\delta = 0$
1. Fisher's Exact Test (chapter 1)
> Technical conditions: Random samples from finite populations or randomized experiment

2. Two-sample z test
> Technical conditions: Independent random samples from processes or large population, and at least 5 successes and 5 failures in each group.
>
> The test statistic $z_0 = \dfrac{\hat{p}_1 - \hat{p}_2}{\sqrt{\hat{p}(1-\hat{p})\left(\dfrac{1}{n_1} + \dfrac{1}{n_2}\right)}}$ where $\hat{p} = \dfrac{X_1 + X_2}{n_1 + n_2}$
>
> can be modeled by the standard normal distribution.

Approximate C% Confidence interval for $\pi_1 - \pi_2$ or for δ
1. Two-sample z interval
> Technical conditions: Independent random samples from processes or large population or randomized experiment, and at least 5 successes and 5 failures in each group.
>
> An approximate C% interval: $(\hat{p}_1 - \hat{p}_2) \pm z^* \sqrt{\dfrac{\hat{p}_1(1-\hat{p}_1)}{n_1} + \dfrac{\hat{p}_2(1-\hat{p}_2)}{n_2}}$
>
> where $-z^*$ is the $(100-C)/2^{\text{th}}$ percentile from the standard normal distribution.
2. Or the Wilson adjustment can be made on the sample proportions and sample sizes first.

Two-sample z-procedures in Minitab
> Choose Stat > Basic Statistics > 2 Proportions.
>
> Remember to convert to the sample counts first and to choose "pooled p" under the Options button.

To test H_0: $\tau = 1$ (equivalent to *log* $\tau = 0$)
> The test statistic $z_0 = (\log(\hat{\tau}))/\sqrt{\dfrac{1}{a} + \dfrac{1}{b} + \dfrac{1}{c} + \dfrac{1}{d}}$
>
> can be modeled by the standard normal distribution for large sample sizes.

> An approximate C% confidence interval for the (population) odds ratio is

$$\exp\left(\log(\hat{\tau}) \pm z^* \sqrt{\frac{1}{a} + \frac{1}{b} + \frac{1}{c} + \frac{1}{d}} \right)$$

Practice:

5-6) Violence Begets Violence (cont.)
Recall from chapter 1 that a researcher searched court records to find 908 individuals who had been victims of abuse as children (11 years or younger). She then found 667 individuals, with similar demographic characteristics, who had not been abused as children. Based on a search through subsequent years of court records, she determined how many in each of these groups became involved in violent crimes (Widom, 1989). The results are shown below:

	Abuse victim	Control
Involved in violent crime	102	53
Not involved in violent crime	806	614

Your earlier analysis with Fisher's Exact test produced a p-value of .015.
(a) Carry out a z test to determine whether abused victims are significantly more likely to be involved in a later violent crime than the control group [*Hint*: Remember to state the hypotheses and check the technical conditions before proceeding]. Use the z interval to estimate the size of the treatment effect. Summarize your conclusions.
(b) How does the p-value of your test compare to the p-value from Fisher's Exact Test?
(c) Calculate a 95% confidence interval for the odds ratio of being involved in a violent crime between these two groups.
(d) Are you willing to conclude that being a victim of abuse as a child causes individuals to be more likely to be violent toward others afterwards? Explain.
(e) Are you willing to generalize these results to all abuse and non-abuse victims?

5-7) AIDS and AZT
A clinical study conducted in the early 90's involved 324 pregnant women who were HIV-positive. Of the 164 women who were randomly assigned to receive the experimental (at the time) drug AZT, 13 had HIV-positive babies. Of the 160 women who were randomly assigned to receive a placebo, 40 had HIV-positive babies.
(a) Is this an observational study or an experiment? Explain.
(b) Identify the observational units, explanatory variable, and response variable.
(c) Organize the data in a 2×2 table, with the explanatory variable in columns.
(d) Calculate the relative risk and the odds ratio of having an HIV-positive baby between the placebo and AZT groups.
(e) Create a segmented bar graph to display the data.
(f) Conduct a z-test of whether the data suggest that AZT is helpful for reducing the rate of HIV-positive babies. Report the hypotheses, test statistic, and p-value. Also write a sentence or two summarizing your conclusion.
(g) Determine and interpret a 99% confidence interval for the odds ratio of having an HIV-positive baby between the AZT and placebo groups as a measure of the true treatment effect.

SECTION 5-3: COMPARING TWO SAMPLES ON A QUANTITATIVE RESPONSE

The previous two sections concerned comparing a categorical response variable between two groups. Now we turn our attention to comparing a quantitative response variable between two groups. Again we start by considering data gathered as independent random samples from two populations, and then we return to considering data collected in a randomized experiment.

Investigation 5-5: NBA Salaries (Minitab Exploration)

The file NBASalaries0203.mtw contains all individual player salaries (in millions of dollars) for the 2002-2003 basketball season as of April 2, 2003 collected by David Dupree of *USA Today* and posted on hoopsworld.com. For now we will focus on comparisons between the Western Conference players and the Eastern Conference players.

(a) Identify the observational units and the variable(s) in this study. Do these data constitute samples or populations?

(b) Produce numerical and graphical summaries of the salaries in each conference using one of the following two approaches:

To use the "stacked" data in cols 2 and 5	To use the "unstacked" data in cols 6 and 7
Choose Graph > Boxplot and select the One Y, With Groups option Enter C2 as the Graph variable and C5 as the categorical variable for grouping. (Under Scale you can transpose the variables.) Choose Stat > Basic Statistics > Display Descriptive Statistics Enter C2 in the Variables box and C5 in the By variables box.	Choose Graph > Boxplot and select the Multiple Y's, Simple option Enter C6 and C7 into the Graph variables box. (Under Scale you can transpose the variables.) Choose Stat > Basic Statistics > Display Descriptive Statistics Enter C6 and C7 in the Variables box.

Write a paragraph describing the behavior of the salaries in each conference. Also compare and contrast the salary distributions between these two populations.

Suppose we did not have access to all of the salaries, but instead had an independent random sample of 20 observations from each conference and calculated the sample mean salary for each conference.

(c) If we were to repeat this sampling process many times, how do you think the distribution of the sample average would behave for each conference? Explain.

You will now simulate this sampling process in Minitab to examine how the distribution of the *differences* in the sample averages behaves. Create a Minitab macro with the following commands:

```
sample 20 c6 c10
sample 20 c7 c11
let c14(k1)=mean(c10)
let c15(k1)=mean(c11)
let c16(k1)=mean(c10)-mean(c11)
let c17(k1)=std(c10)
let c18(k1)=std(c11)
let k1=k1+1
```

Remember to initialize the counter `k1` and then execute this macro 1000 times. When the macro has finished executing, make sure to name C14 – C18 appropriately.

(d) Produce numerical and graphical summaries of the sample means obtained for each conference (C14 and C15). Do these distributions behave as you predicted above? Explain.

(e) Create numerical and graphical summaries of the differences in the sample means (C16). Be sure to report the mean and standard deviation of these differences. Describe the behavior of this distribution. Does it behave as you expected? Explain.

(f) Create a normal probability plot of the differences in C16. Do you think these differences in sample means are reasonably modeled by a normal distribution?

(g) Using the probability rule discussed on page 5-6, derive a general expression for the expected value of the difference the sample means between two independent random samples, $E(\bar{X} - \bar{Y})$.

(h) Using the probability rule discussed on page 5-6, derive a general expression for the standard deviation of the difference in samples means between two independent random samples, $SD(\bar{X} - \bar{Y})$.

(i) Calculate the theoretical mean and standard deviation of the sampling distribution of $\bar{X}_{east}$- $\bar{X}_{west}$ for the NBA populations where $\mu_{east}=$ 3.580, $\mu_{west}=$ 3.960, $\sigma_{east}=$ 3.773, $\sigma_{west}=$ 4.396. How do these theoretical values compare to the simulated values?

(j) If we do not know the population means μ_1 and μ_2, then we probably don't know the population standard deviations σ_1 and σ_2 either. Suggest an expression for the standard error to approximate $\sqrt{\dfrac{\sigma_1^2}{n_1} + \dfrac{\sigma_2^2}{n_2}}$, the standard deviation of the difference in sample means.

(k) If we standardize the difference in sample means with this standard error to obtain

$$\frac{(\bar{X}_1 - \bar{X}_2) - (\mu_1 - \mu_2)}{\sqrt{\dfrac{s_1^2}{n_1} + \dfrac{s_2^2}{n_2}}}$$

what type of distribution do you think these standardized values will follow? Explain.

(l) In Minitab, create these standardized values for your 1000 simulated pairs of samples:

```
MTB> let c20 = (c16-mean(c6)+mean(c7))/sqrt(c17**2/20+c18**2/20)
```

Describe the distribution and use a normal probability plot to decide whether these sample values appear to be reasonably modeled with a normal or t distribution.

(m) Repeat (c)-(f) but change your macro to store the *medians* instead of the means. Do the differences in the sample medians in C16 appear to follow a normal distribution? Explain.

Discussion: Even though the population distributions of salaries are both sharply skewed to the right, a sample size of 20 is reasonably sufficient for the sampling distributions of the sample *mean* for each league to look relatively normal. Since the two individual sampling distributions roughly follow a normal distribution, so does the distribution of their differences. Furthermore, the mean of the differences will equal the difference in the population means and the variance of the differences will be equal to the sum of the variances. Moreover, because the distribution of $\bar{X}_1 - \bar{X}_2$ is approximately normal, the statistic $\dfrac{(\bar{X}_1 - \bar{X}_2) - (\mu_1 - \mu_2)}{\sqrt{\dfrac{s_1^2}{n_1} + \dfrac{s_2^2}{n_2}}}$ will approximately follow a t-distribution. These ideas, including the degrees of freedom, are stated more formally in the following probability detour. Unfortunately, we don't have as simple a model when looking at the difference in medians or other statistics. In these situations we can use other techniques such as bootstrapping which will be considered in a later activity.

Probability Detour

A theorem similar to the Central Limit Theorem indicates that if we have two infinite, normally distributed populations, with means μ_1 and μ_2 and standard deviations σ_1 and σ_2, and we take independent random samples from each population, the sampling distribution of $\overline{X}_1 - \overline{X}_2$ will follow a normal distribution with mean $\mu_1 - \mu_2$ and standard deviation $\sqrt{\dfrac{\sigma_1^2}{n_1} + \dfrac{\sigma_2^2}{n_2}}$.

This is a nice theoretical result, but we seldom meet these conditions exactly.

- The first issue is that if we don't know μ_1 and μ_2, then we most likely don't know σ_1 and σ_2 either. Luckily, the t distribution comes to our rescue again. If we substitute the sample standard deviations, s_1 and s_2, and standardize using this *standard error* for $\overline{X}_1 - \overline{X}_2$, we obtain the statistic $t = [(\overline{X}_1 - \overline{X}_2) - (\mu_1 - \mu_2)] / \sqrt{\dfrac{s_1^2}{n_1} + \dfrac{s_2^2}{n_2}}$, which approximately follows a t distribution. The degrees of freedom can be conservatively approximated by taking $\min(n_1-1, n_2-1)$ or statistical software will approximate them from the data (and may not even obtain an integer).

- The second issue is that we don't usually have normally distributed populations. As you saw in the above Minitab Exploration, the sampling distribution of the differences will still be approximately normal as long as the sample sizes are large. How large the sample sizes need to be of course depends on the degree of nonnormality in the population distributions, but in many practical situations, sample sizes of about 20 are sufficient. The approximation is even better if the populations have similar shapes and the sample sizes are similar, as they were with the NBA data. You need to explore the graphical displays of the two samples to help you make this judgment call. In fact, the t procedures are fairly *robust* to departures from the normality condition.

- The third issue is that we don't typically have infinite populations. Similar to the binomial approximation to the hypergeometric distribution, we will consider the populations large enough if they are more than 20 times the size of the samples. Most important is that the samples are randomly selected and that they can be considered independent of each other. This will be true if we employ a stratified sampling procedure but in practice this procedure will generally be used as long as the samples are not dependent (e.g., repeated observations on same individuals).

- Another issue to keep in mind is whether a comparison of centers is appropriate in the first place. To compare only the population centers we are implicitly assuming that the shapes and standard deviations are already similar (if the std devs were quite different, then comparing the centers may not be meaningful). In fact, if we are willing to assume that the population standard deviations are equal, we can simplify the calculation of the t-statistic even further. If we assume $\sigma_1 = \sigma_2 = \sigma$, then, when the hypothesis of equal population means is assumed, our best estimate of σ is a weighted average of the sample variances:

$$s_p = \sqrt{\dfrac{(n_1 - 1)s_1^2 + (n_2 - 1)s_2^2}{n_1 + n_2 - 2}} \ . \ \text{Then we use SE}(\overline{X}_1 - \overline{X}_2) = s_p \sqrt{\dfrac{1}{n_1} + \dfrac{1}{n_2}}.$$

Standardizing with this standard error leads to a t distribution with degrees of freedom equal to $n_1 + n_2 - 2$ and is referred to as the *pooled t-test*. However, it can be difficult to ascertain whether the population standard deviations are truly equal and the advantages in doing so are not large. Thus we recommend using the unpooled t procedures in sampling situations.

Practice:

5-8) NBA Salaries (cont.)

Return to your simulation macro for the difference in sample means.

(a) Repeat the above simulation analysis using sample sizes of 5 for each conference instead of 20. Examine both the empirical sampling distribution of the difference in means and the t statistics. Are the sample sizes large enough for the shape of the sampling distribution of the difference in means to be symmetric? For the distribution of t statistics?

(b) Use the expression derived above to calculate the theoretical value of SD($\bar{X} - \bar{Y}$). Then compare this to the standard deviation of your simulated sample means.

(c) Repeat the above simulation using a sample size of 5 for one conference and a sample size of 20 for the other, instead of equal sample sizes. Examine both the empirical sampling distribution of the difference in means and the t statistics. Are the sample sizes large enough for the shape of the sampling distribution of the difference in means to be symmetric? The t statistics?

5-9) Body Temperatures (cont.)

Recall the body temperatures data (`temps.mtw`). The gender of each individual was also recorded and stored in column 2.

(a) State the null and alternative hypotheses for testing whether the average male body temperature differs from the average female body temperature.

(b) Treat the data as independent random samples of men and women. Calculate the test statistic $t_0 = \dfrac{(\bar{x}_1 - \bar{x}_2) - 0}{\sqrt{\dfrac{s_1^2}{n_1} + \dfrac{s_2^2}{n_2}}}$. Interpret this value.

(c) Approximate the p-value for this test statistic from a t distribution with $min(n_1-1, n_2-1)$ degrees of freedom. Would you reject or fail to reject the null hypothesis at the 5% level? What does this tell you about male and female body temperatures?

(d) Calculate and interpret a 95% confidence interval to estimate the difference in the population means using these data. [*Hint*: Follow the convention of: *point estimate* $\pm$ (*critical value*)×(*standard error*). Again use $min(n_1-1, n_2-1)$ to approximate the degrees of freedom for the t critical value.]

(e) Verify these calculations in Minitab by choosing Stat > Basic Statistics > 2-Sample t.

(f) Comment on the validity of the t-procedures for these data.

Investigation 5-6: Handedness and Life Expectancy

Psychologist Stanley Coren has conducted several studies investigating the life expectancy of left-handers compared to right-handers, believing that the stress of being left-handed in a right-handed world leads to earlier deaths among the left-handers. In one study Coren and Halpern (1991) sent surveys to thousands of next-of-kin of recently deceased southern Californians and asked whether the person had been right-handed or left-handed. They were very careful in how they collected their data. First, they consulted a bereavement counselor who suggested that they not contact anyone unless at least 9 months had passed since the death. The counselor also suggested that they make the contact as gentle as possible and not follow up or press people for responses. The researchers also decided that they would not contact next of kin if the death had been a result of murder or suicide or if the deceased was a child age 6 or younger. They received 987 replies and found that the average age of right-handed people at death was 75 years and for left-handed people it was 66 years.

(a) Is this an observational study or an experiment? Is it retrospective or prospective? Based on your responses, how will this affect the scope of conclusions you will be able to draw from this study?

(b) Did the researchers take a random sample of left-handers and an independent random sample of right-handers? Is it reasonable to consider these samples independent?

> Even though the researchers did not take a random sample of left-handers and a separate random sample of right-handers, we are still willing to consider these samples as independent since the results for one group will have no effect on the results for the other group.

(c) The summary of the study above does not report the standard deviations and sample sizes of each group. Is this important information? Explain.

(d) Let μ_L represent the mean lifetime for the population of left-handers and let μ_R represent the mean lifetime for the population of right-handers. State the null and alternative hypotheses for Coren and Halpern's study, both with symbols and in words.

(e) For the scenarios in the following table, which do you believe will result in a lower p-value, i.e., stronger evidence against the null hypothesis? Explain briefly.

Scenario 2 vs. 3

Scenario 3 vs. 4

(f) Use Minitab (version 14) to carry out the two-sample t tests for these hypotheses, using the sample sizes, sample means, and sample standard deviations given in the table above in the summarized data section of the 2-Sample t command box (Stat>Basic Statistics>2-Sample t). For each of these five scenarios, report the resulting test statistic, p-value, and whether or not the difference is statistically significant with $\alpha=.10$.

Scenario		Sample sizes	Sample means	Sample SDs	t-statistic	p-value	Significant at 10% level?
1	left	99 (10% of 987)	66	15			
	right	888	75	15			
2	left	50 (5% of 987)	66	15			
	right	937	75	15			
3	left	50 (5% of 987)	66	25			
	right	937	75	25			
4	left	10 (1% of 987)	66	25			
	right	977	75	25			
5	left	99 (10% of 987)	66	50			
	right	888	75	50			

Write a paragraph summarizing the effects of the sample sizes and the sample standard deviations on the values of the t-statistic and p-value.

(g) Considering the five scenarios of sample sizes and standard deviations used in the table, which combination do you think is most realistic for this study of left- and right-handers' lifetimes? Explain, based on the context for these data.

(h) Even without knowing the exact values of the sample sizes and the standard deviations, does the difference in mean age at death observed by the researchers appear to be statistically significant? Explain.

(i) Select the most realistic of the five scenarios presented in the table, and construct a 95% confidence interval for the difference in mean ages at death between the two groups. Also write a sentence or two interpreting the interval.

(j) Part of the motivation for this research was an earlier study by Porac and Coren (1981) that surveyed 5147 men and women of all ages in North America. They found that 15% of ten-year-olds were left-handed, compared to only 5% of fifty-year-olds and less than 1% of eighty-year-olds. At the age of 85, right-handers outnumbered left-handers by a margin of 200 to 1. Suggest another explanation for these puzzling findings that actually provides a counter-argument to the conclusion from the 1991 Coren and Halpern study that left-handers tend to die younger than right-handers.

(k) Explain why it is not feasible to conduct a randomized, comparative experiment to investigate whether going through life as a left-hander causes a person to tend to die at a younger age.

Study Conclusions: The difference in the sample means does appear to be statistically significant for all reasonable choices of the sample sizes and the sample standard deviations. However, there are numerous cautions to heed when drawing conclusions from such a study. This was a retrospective study and in fact the researchers reported that they tended to hear from the left-handed relatives more often. Still they only heard from fewer than half of the families contacted, so there is strong likelihood of bias in their sampling method. There is also no information given about the proportion of left-handers in the two southern California counties studied or the average ages of their residents now. In particular, one explanation for the lower percentage of left-handers among the elderly is not that they have died younger but that it used to be quite common practice to strongly encourage left-handed children to switch to being right-handed. Nowadays, that is less common (in fact many athletes love having this advantage!), and so there is a higher percentage of left-handers among younger age groups. This helps to explain why the left-handers who had died would tend to be younger. In other words, maybe the average age difference between *living* left-handers and right-handers is also nine years.

These studies have actually become hot topics for debate as some other studies have not been able to replicate Coren and Halpren's results. Other prospective longitudinal studies in the United States (Marks and Williams, 1991; Wolf, D'Agostino, and Cobb, 1991) have not found a significant difference in age at death. Still others have found connections between handedness and accident rates, lower birth rates, cancer, alcohol misuse, smoking, and schizophrenia. Alas, we don't see any randomized, comparative experiments answering these questions soon!

Discussion: As you first saw in chapter 2, you should have found that larger variability in each sample produces a larger p-value and therefore less convincing evidence that the sample means differ significantly. You should also have found that a bigger discrepancy in sample sizes between the two groups produces a larger p-value and therefore less convincing evidence that the sample means differ significantly.

Practice:

5-10) Left-Handed Advantages?

Noroozian, Lotfi, Gassemzadeh, Emani, and Mehrabi (2002) compared the acceptance rate of left-handers with that of right-handers in the College Entrance Examination (CEE) for the national universities in Iran. About one million Iranian high school graduates take part each year in the CEE. An entrance exam score is obtained for each participant, which has a mean of 5000 and a standard deviation of 100. A comprehensive list of all the participants between 1993-1997 was obtained and 10,000 were chosen randomly from each year. Hand preference was exclusively defined as writing preference. The distribution of left-handers and the distribution of right-handers did not differ significantly with respect to gender. Of the 47,854 right-handers, the mean score on the College Entrance Exam was 5020 with standard deviation 718. Of the 3,398 left-handers, the mean CEE score was 5060 with standard deviation 720.

(a) Is it appropriate to apply the two-sample t-procedures to these sample data, or do you not have enough information to decide? Explain.

(b) Is this a statistically significant difference in the mean CEE score between the population of right-handers and the population of left-handers?

(c) Compute a 95% confidence interval for the difference in mean score between left-handed population and the right-handed population.

(d) Explain how this difference may be considered statistically significant but not practically significant. What is the cause for this?

Investigation 5-7: Comparison Shopping (cont.)

Recall the goal of comparing prices between two different grocery stores. The file shopping99.mtw contains results for Lucky's, a "low-price grocery store" (at least their ads say so) and Scolari's, a local, privately owned chain. The data were collected by statistics students in the Fall of 1999 using a sample that was constructed by randomly selecting products from shopping receipts of numerous students. We earlier removed any products that were not exactly matched at both stores; remove these observations again.

(a) Use Minitab to construct graphical and numerical summaries for comparing the prices at these two stores. Comment on what your summaries reveal.

(b) State the hypotheses for testing whether "the low price leader" tends to have lower prices, comment on the technical conditions, and carry out a two-sample t-test to examine whether the difference between the two sample means is statistical significant at the $\alpha=.05$ level. Summarize and explain your conclusion.

(c) Explain why this two-sample t-test is inappropriate here.

(d) Explain the advantage of finding the same item at both stores instead of taking a separate random sample of items from each store.

The chief difficulty with the above analysis is that the two samples are not independent but clearly paired. In fact, recall that the students were specifically instructed to find the same item at both stores. If the identical product could not be found, that product was dropped from the list. This isolates the "store effect." If the students have found the same exact item each time, then the only reason the price should vary is due to a store-to-store difference.

Recall that with paired data, it makes sense to analyze the *differences* and determine whether the mean difference is far from zero. In chapter 4 you did so using bootstrapping. When the sample size is large or the population of differences is roughly normal, we can apply the *t*-distribution instead.

(e) Use Minitab to calculate the differences in prices of these items (`let c4=c2-c3`), and then produce numerical and graphical summaries of these differences and discuss what they reveal.

(f) State the hypotheses for testing whether the low price leader tends to have lower prices, check the technical conditions, sketch the standardized sampling distribution, and carry out a one-sample *t*-test to investigate whether the sample mean of the *differences* in price is significantly less than zero at the $\alpha=.05$ level. Summarize and explain your conclusion.

(i) How do the test statistic and p-value from the one-sample *t*-test on the differences compare to their counterparts from the two-sample *t*-test? Have they changed much?

(j) To see why the test statistic has changed so much, calculate the means and standard deviations of these three variables (MTB> describe c2-c4) and record them in the table below. Use these statistics to suggest why the test statistic based on the sample of differences is so much larger than that based on independent samples. [*Hint*: Think about whether the numerator or denominator in the test statistic calculation has changed.]

	Lucky's	Scolari's	Difference
mean			
standard deviation			

(k) Use Minitab to produce a 90% confidence interval for the population mean price difference for these two stores. Interpret this interval, and comment on the importance of whether this interval includes the value zero. Do you consider this price difference practically significant? Explain.

(l) Conduct a sign test to assess whether Lucky's has significantly more products at a lower price than Scolari's based on these data.

Study Conclusions: Judging from the store's reputations and advertisements, we have tested whether these data provide evidence that Lucky's has lower prices (a one-sided test). One product (milk) was removed from the data set since it was not exactly the same size at both stores. After doing this, a graph of the differences appears reasonably symmetric. There are a few outliers, but we don't have reason to doubt the accuracy of their measurements. The sample size is almost above 30 so since the sample is reasonably symmetric we consider the one-sample *t*-procedures to be applicable here. The one-sided paired *t*-test reveals a (barely!) statistically significant (at the 5% level, p-value = .046) mean price difference between Lucky's and Scolari's. However, this difference appears to be dramatically affected by a few items that had much cheaper prices at Lucky's—Breyer's vanilla frozen yogurt, and Crest regular toothpaste. The sign test, which only counts how many prices were more expensive at Scolari's, produces a larger p-value of .19, and we would not consider Lucky's to have significantly more products at a lower price than Scolari's. Of course, part of the explanation for the larger p-value with the sign test is that this test ignores information about the magnitudes of the price differences and so is less powerful (i.e., less likely to detect a difference when there really is one) than the *t*-test.

Discussion: The *paired t-test* is more appropriate here since we have dependent samples. It is also much more informative to directly compare the same product at both stores. The key advantage of *pairing* is that it *reduces variability*. The average price difference is about 16 cents. This value is not statistically significant when compared to the standard deviations of prices at the two stores (1.66 and 1.74), but is significant when compared to the smaller standard deviation of the differences (.408). By controlling for the variation in prices from product to product, we are better able to see a difference between the stores. Still, this difference is not huge. We are 90% confident that the average price difference is between about 3 cents and 29 cents (in favor of Lucky's) per item. A shopper could consider how many items they plan to buy before deciding if this average price difference is worth traveling to the store that is further away.

Practice:

5-11) Walk this Way

If you want to study whether people walk faster with their arms held at their sides or with their arms swinging up and down, explain why a paired design would be helpful and how you might implement the study. [*Hint*: Where would randomization come into play?]

5-12) Exam Performance

Suppose that you want to compare students' performances on the first two exams in a course.

(a) Would it make more sense to use a paired test or an independent sample test? Explain.

(b) For the summary data provided below, calculate the paired t-statistic and p-value, and also the independent-samples t-statistic and p-value. Does pairing appear to have been useful in this situation? Explain.

exam1:	$n_1=12$	$\bar{x}_1=86.4$	$s_1=9.5$
exam2:	$n_2=12$	$\bar{x}_2=83.3$	$s_2=12.3$
differences:	$n_d=12$	$\bar{x}_d=3.2$	$s_d=4.5$

(c) Repeat (b) for the summary data provided below. [*Hint*: If you pay close attention, you can avoid duplicating work.]

exam3:	$n_3=12$	$\bar{x}_3=86.4$	$s_3=9.5$
exam4:	$n_4=12$	$\bar{x}_4=83.3$	$s_4=12.3$
differences:	$n_d=12$	$\bar{x}_d=3.2$	$s_d=18.0$

(d) Explain why pairing is so effective in one case and not in the other. You may want to speculate about what else might be true about how students' exam performance is related across the exams.

5-13) Schizophrenic Twins (cont.)

Recall the study of the volumes of the hippocampus brain regions of monozygotic twins that are discordant for schizophrenia from Investigation 3-13 (`Hippocampus.mtw`).

(a) Carry out a two-sample t-test using these data. What conclusion would you draw about whether the mean hippocampus volumes differ between those affected and those unaffected by schizophrenia?

(b) Explain why this test is inappropriate in light of the way the data were collected.

(c) Compare these results to those from the one-sample t-test from the end of Section 4-4. Does the pairing appear to have been effective? Explain.

SECTION 5-4: RANDOMIZED EXPERIMENTS REVISISTED

In Section 5-1, we began by considering independent random samples from two populations, and comparing them on a categorical response variable. We found in Section 5-2 that the same two-sample z-procedure applied to data gathered in a randomized comparative experiment. Then in Section 5-3 we switched back to considering data from independent random samples, this time focusing on a quantitative response variable. We found that inference procedures that theoretically apply for random samples from infinite and normally distributed populations seem to also work well in practice with finite and nonnormal populations when the sample sizes are large. Now you will see that these t distributions are also a useful approximation to the randomization distributions discussed in chapter 2 for comparing quantitative responses from a randomized comparative experiment.

Investigation 5-8: Sleep Deprivation (cont.)

Recall the sleep deprivation study of chapter 2 that examined the performance of subjects who had been deprived of sleep on a visual discrimination task.

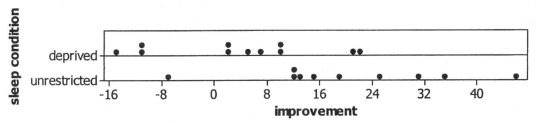

Sleep group	Sample size	Mean improvement	Median improvement	Standard Deviation
Deprived	11	3.90	4.50	12.17
Unrestricted	10	19.82	16.55	14.73

In chapter 2 we used a randomization test to assess the significance of the difference in the mean performance of the two groups. We tested whether the sleep deprived group tended to perform worse on the exam than the unrestricted sleep group. The following graphs display the results for 1000 randomizations assuming no treatment effect (H_0: $\delta=0$).

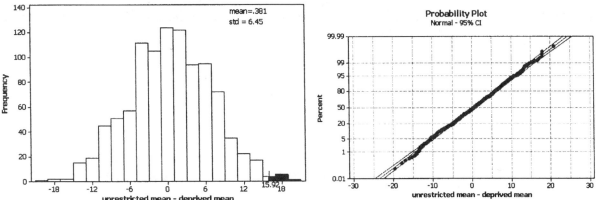

The above output yields an empirical estimate of the p-value around .008, giving us strong evidence that sleep deprivation leads to a decrease in average exam performance ($\delta<0$). We could improve this estimate by simulating more randomizations. We saw in chapter 2 that if we carried out all possible random assignments, the exact p-value was .0072.

In many situations, the randomization distribution is well approximated by the t distribution, allowing us to estimate the p-value from this distribution instead of from the randomization distribution. The empirical randomization distribution pictured above is roughly symmetric but we need to estimate the standard deviation, so it is reasonable to believe that the standardized values follow a t distribution. Since we are considering the 21 observations as the population and the two groups are being formed from the same population, we will consider the *pooled two-sample t-test*.

(a) Carry out a *pooled* t-test to compare the mean performance of the sleep deprived and the unrestricted groups. Report the hypotheses, test statistic, and the p-value. How does the p-value approximation compare to that from the randomization simulation? [*Hint*: If you are conducting this test by hand, remember that the pooled sample standard deviation can be found from

$$s_p = \sqrt{\frac{(n_1 - 1)s_1^2 + (n_2 - 1)s_2^2}{n_1 + n_2 - 2}} \; .]$$

The following histogram applies to the same simulation as above but having calculated the *pooled t-statistic* for each repetition of the randomization process.

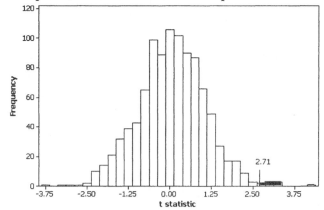

We observed 8 of the 1000 t-statistics exceeding the observed value in the study, 2.71. This corresponds to an empirical p-value of .008 -- very close to that reported by Minitab.

The graph below shows the results for all 352, 716 randomizations possible with the t distribution overlaid.

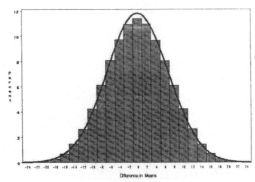

This investigation shows that the *t* distribution provides a suitable approximation, especially when it would not be feasible to list all possible random assignments and count how many have a test statistic more extreme than the one observed. In addition, the *t* procedure provides an approximate confidence interval for the true treatment effect. To do so with the randomization procedure we would have to carry out a randomization procedure for each conjectured value in the interval and see if the resulting two-sided empirical p-value exceeds α.

(b) Calculate and interpret a 95% confidence interval for the treatment effect δ based on this study. [*Hint*: Use a pooled *t*-procedure.]

(c) Would you be willing to generalize these results to a larger population of subjects? Explain.

Discussion: The pooled *t*-test provides another way of assessing the treatment effect when the sample sizes are reasonably large by providing an approximate p-value and an approximate confidence interval for the treatment effect. Since the randomization was in the assignment of subjects to conditions, we do need to be cautious in generalizing the results from this study to any larger population.

In using the pooled *t*-procedures we additionally need to assess whether the groups have similar variability. These procedures are reasonably robust against both nonnormality and unequal standard deviations when the samples sizes are similar, but can be quite sensitive to unequal standard deviations otherwise and should be used with caution unless the sample sizes are large.

Practice:

5-14) Low-Carb Diets
"Low-carb" diets have received much attention recently, but there have not been many studies on their efficacy. A recent study by Foster el al. was reported in the *New England Journal of Medicine* (May, 2003). The researchers randomly assigned 63 obese men and women to either a low-carbohydrate, high-protein, high-fat (Atkins) diet or a low-calorie, high-carbohydrate, low-fat (conventional) diet. The mean amount of weight lost, as percent of body weight, after 3 months, 6 months and 12 months are shown in the table below. (The baseline weight was carried forward in the case of missing values.)

Time	Diet	Sample size	Mean	Standard deviation
3 months	Low-carb	33	6.8	5.0
	Conventional	30	2.7	3.7
6 months	Low-carb	33	7.0	6.5
	Conventional	30	3.2	5.6
12 months	Low-carb	33	4.4	6.7
	Conventional	30	2.5	6.3

(a) Is this an observational study or an experiment? Explain.
(b) Identify the explanatory and response variables.
(c) Use Minitab (remember the "summarized data" option in the Stat> Basic statistics> 2-Sample t command box) to test whether the sample mean weight losses differ significantly between the two diets at the 3-month point. Report the test statistic, p-value, and your test decision at the $\alpha = .05$ significance level.
(d) Repeat (c) for the weight losses at the 6-month point and at the 12-month point.
(e) Report the 95% confidence intervals (as produced by Minitab) for the difference in mean weight loss between the two diets at each time point. Comment on how they change over time.
(f) Write a paragraph summarizing the results of this study, being sure to comment on the scope of conclusions you can draw (Can you generalize these results to a larger population? Can you draw a cause and effect conclusion? Justify your decisions.).

5-15) Cloud Seeding (cont.)
Recall the cloud seeding experiment from chapter 2. There you looked at numerical and graphical summaries to see if there was evidence that seeding clouds increased the amount of rainfall. Open the CloudSeeding.mtw worksheet.
(a) Would it be appropriate to carry out a two-sample *t*-test to assess whether the treatment effect is statistically significant? Explain.
When data are strongly skewed, another alternative is to transform the data to a scale where the distribution of the variable is more symmetric.
(b) Take the natural log of each column (MTB> let c5=log(c1)) and produce graphical and numerical summaries of the distribution of the log rainfall amounts. Are these distributions now fairly symmetric?
(c) How has the transformation affected the comparison of the standard deviations of each group? Explain why this is also advantageous.
(d) Carry out a two-sample *t*-test on the transformed variable to compare the two groups. Write a paragraph summarizing your conclusions.

Summary of two-sample *t* procedures

To test H_0: $\mu_1 - \mu_2$ = *hypothesized difference* or δ = *hypothesized treatment effect*

The test statistic $t_0 = \dfrac{(\bar{x}_1 - \bar{x}_2) - hypothesized\ difference}{\sqrt{\dfrac{s_1^2}{n_1} + \dfrac{s_2^2}{n_2}}}$ can be modeled by a *t* distribution.

Approximate C% Confidence Interval for $\mu_1 - \mu_2$ or δ:

Interval: $(\bar{x}_1 - \bar{x}_2) \pm t^{*}\ \mathrm{SE}(\bar{X}_1 - \bar{X}_2)$

where $\mathrm{SE}(\bar{X}_1 - \bar{X}_2) = \sqrt{\dfrac{s_1^2}{n_1} + \dfrac{s_2^2}{n_2}}$ and $-t^{*}$ is the $(100\text{-}C)/2^{\text{th}}$ percentile of the *t* distribution.

The calculation of the degrees of freedom for these two procedures is complicated and will be determined by Minitab. A conservative approximation of the degrees of freedom is the minimum value of n_1-1 and n_2-1.

Technical conditions: These procedures are valid as long as the data can be considered independent random samples from large populations or from a randomized experiment; either the population distributions must both be approximately normal or the sample sizes need to be large, e.g., both at least 20. The *t*-procedures are fairly robust to this second condition.

Two-sample *t*-procedures in Minitab

Choose Stat > Basic Statistics > 2-Sample t where you can use the "stacked" data, the "unstacked" data or the summary statistics.

Choosing "assume equal variances" performs the pooled *t*-procedures, but we recommend using the unpooled procedures when comparing population means. If analyzing data from a randomized experiment, you may choose to pool, but you should first check that the standard deviations are similar. A convention is to check whether the ratio of the larger to smaller standard deviation is less than 2, $(s_{max}/s_{min} < 2)$.

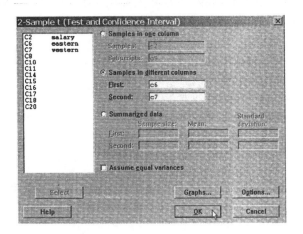

SECTION 5-5: OTHER STATISTICS

Investigation 5-9: Heart Transplants and Survival

Heart transplantation is no longer considered an experimental treatment but is used on thousands with end-stage organ disease. More than 22,000 Americans receive an organ transplant each year. Stanford University has long been recognized as the pioneering center for heart transplants and carried out the first adult heart transplantation in the U.S. in 1968. In the mid-1970's, data were examined to assess the effect of transplantation on survival. Crowley and Hu (1977) examined the data for the patients enrolled between September, 1967, and March, 1974. After a patient is enrolled, a donor heart, matched on blood type, is then sought. This wait could last anywhere from a few days to almost a year. Some patients die before a suitable heart is found. When a match is found, there could be as many as seven patients waiting. It is generally thought that no serious bias has been introduced by the selection of patients for the transplant operation. In fact, physicians believe that, if anything, less hardy patients tend to receive hearts preferentially over hardier ones. In the data set you will examine, 69 patients received a transplant and 34 did not. The researchers recorded the number of days they survived from the time of enrollment until death or until the study ended. These data can be found in `transplants.mtw`.

(a) Produce numerical and graphical summaries to compare the survival times of the transplant patients with those that did not receive a transplant. Write a paragraph summarizing the results.

(b) Comment on the usefulness of comparing the *mean* survival times of these two distributions.

Suppose we consider the two samples to be representative of larger populations of transplant and non-transplant patients. Since survival times are often heavily skewed to the right, and are sometimes *truncated* at the conclusion of the study for subjects who are still alive, researchers may choose to compare the medians instead of the means. Since the sampling distribution for the distribution of medians is not well modeled by the normal or *t* distributions, we can again consider *bootstrapping* to mimic the random sampling process and obtain an estimate of the sampling variability of the difference in sample medians.

(c) What is the observed difference in the median survival time of the patients who received a heart transplant and the patients who did not receive a heart transplant?

(d) Use a Minitab macro to resample, with replacement, from each treatment group:

```
sample 69 c3 c5;
replace.
sample 34 c4 c6;
replace.
let c7(k1)= median(c5)-median(c6)
let k1=k1+1
```

Initialize k1 and execute the macro 1000 times. Describe the shape, center, and spread of the resulting bootstrap distribution of the differences in the medians. Does a normal model appear appropriate for these differences?

Despite the lack of normality, the standard deviation of the bootstrap distribution still provides an estimate of the sampling variability in these differences in the sample medians.

(e) Based on the above simulation results, what number might you report as a standard error of the difference in sample medians?

We can construct an approximate 95% *bootstrap interval* by determining the 2.5^{th} and 97.5^{th} percentiles of the bootstrap distribution.

(f) The 2.5^{th} percentile corresponds to which of the 1000 bootstrap differences? The 97.5^{th} percentile corresponds to which of the 1000 bootstrap differences?

(g) Sort the entries in C7, storing the results in C8 (MTB> sort c7 c8). Examine the results in C8 to find the 2.5^{th} and 97.5^{th} percentiles. [*Hint:* You can scroll through the Data Window, or to see the value in the kth row of a column, you can type

```
MTB> let k1=c8(k)        substituting the value for k
MTB> print k1                      ]
```

(h) Does the interval in (g) contain 0? If not, does the interval lie above or below zero? What does this tell you about the significance of the difference in population median survival time for heart transplant patients and non-transplant patients?

(i) Determine a 90% bootstrap interval based on your simulation results. How does it compare to the 95% bootstrap interval?

Discussion: We must use extreme care in constructing the bootstrap interval. Since the sampling distribution is skewed, it may not be reasonable to use symmetric percentiles as we did above. There are other techniques for calculating bootstrap intervals for skewed distributions. Many software packages will construct such intervals (e.g., the bootstrap bias-corrected accelerated BCa method and the bootstrap tilting method). However, the BCa is not always recommended for medians.

The percentile bootstrap illustrated here is one of the simplest bootstrap confidence intervals methods, but it is may not be the best method in all applications.

Suppose we did not want to assume these data were representative samples from independent populations, but we still wanted to explore whether the size of the treatment effect could have plausibly arisen by chance. We can then apply the analysis methods from chapter 2 to explore the significance of this result. In the following analysis, we will continue to use the difference in medians as the statistic of interest.

(j) Use Minitab to create the randomization distribution for the difference in medians. Your macro should contain the following commands:

```
sample 103 c2 c10
unstack c1 c11 c12;
subs c10.
let c13(k1)=median(c11)-median(c12)
let k1=k1+1
```

Perform at least 1000 repetitions of the randomization. Produce numerical and graphical summaries of the resulting empirical randomization distribution and describe the distribution.

(k) How often do you find a difference in group medians in the empirical randomization distribution that is at least as extreme as that obtained by the researchers?

(l) What conclusion can you draw about the effectiveness of the heart transplants on the lifespan of heart patients based on the empirical p-value in (k) and the type of study conducted here?

Study Conclusions: This data set has been repeatedly analyzed by researchers, namely in an attempt to control for many other lurking variables such as age, prior operations, and other health conditions. Newer, more involved statistical methods (especially in the area of survival analysis) have allowed for more sophisticated analyses of these data. Still, our basic analysis of the data has shown that there is strong evidence of a difference in median survival times between these groups. We considered the median lifetime as more appropriate than the mean lifetime due to the skewness of the survival times and the "censoring" of lifetimes for patients that are still alive at the conclusion of the study. It turns out that the evidence of the benefits of heart transplants is made even stronger when the response variable considers the "quality of life" of the patient in addition to survival time.

Practice:

5-16) NBA Salaries revisited

(a) Open the NBASalaries0203.mtw worksheet. Take a random sample of 10 salaries from the eastern conference and a random sample of 20 salaries from the western conference. Examine your sample results. If you began with these two samples, would it be reasonable to apply a two-sample t procedure to compare the population means? Explain.

(b) Use a Minitab macro to resample, with replacement, from each league and compute the difference in sample *medians*. Execute the macro 1000 times. Describe the shape, center, and spread of the resulting bootstrap distribution of the difference in medians.

(c) Use the method in Investigation 5-8 to calculate a 95% bootstrap interval. Does your interval succeed in capturing the true difference in the population medians ($2.154-$2.437 = -$2.83)?

(d) You are told the method in (c) leads to "95% confidence" intervals. Suggest a way for assessing the accuracy of this statement.

5-17) Heart Transplants (cont.)

(a) Transform the survival times by taking the log (MTB> let c5=log(c1)). Produce numerical and graphical summaries for comparing the log-survival time between the transplant and non-transplant patients. Would it be reasonable to apply the t-procedures to these data?

(b) Calculate and interpret a 95% t confidence interval for the difference in mean-log-survival times between these two groups.

(c) Exponentiate the endpoints of this interval to obtain a confidence interval for the *ratio* of the medians in the two groups. Explain why this works.

SUMMARY

In this chapter you have explored various techniques for comparing two populations, assuming you have selected independent simple random samples from each population. You also saw that these techniques provide large-sample approximations to the randomization tests carried out in chapters 1 and 2.

When comparing two proportions, you should consider the following options:
- Fisher's Exact Test
- Two-sample z-test and interval for the difference in proportions (with larger sample sizes)
- Confidence interval for odds ratio (when the proportions of success are near 0 or 1)
For the two-sample z confidence intervals, the Wilson adjustment helps maintain the nominal confidence level for smaller sample sizes (with larger sample sizes, the intervals will be essentially the same). Many researchers will always work with the odds ratio instead of the difference in proportions as it is the more relevant parameter with extreme success rates and can be utilized with case-control studies when the difference in proportions is not relevant.

When comparing two means, you should consider the following options;
- Empirical randomization test and/or bootstrapping
- Two-sample t-test and interval for the difference in means (with normal populations or larger sample sizes).
The advantage to the computer-intensive procedures is they allow us to work with many types of statistics (such as median and trimmed mean) where we do not have theoretical results on the large sample behavior of the sampling distribution.

Some of the main themes you saw again in this chapter are that the sample size is important in assessing statistical significance, as is the amount of variation in the samples for quantitative data. It is always important to consider the technical conditions of an inference procedure before applying the procedure (e.g., make sure you really have two independent samples). If data are paired, then you should consider one-sample procedures on the differences, increasing the efficiency of the study. You were also again reminded of the importance of considering whether there is random sampling and/or randomization in the study design and how that impacts the scope of conclusions that you can draw from your study.

SUMMARY OF PROCEDURES FOR COMPARING POPULATION PROPORTIONS OR ANALYZING TREATMENT EFFECTS

	Binary Data ($\pi_1-\pi_2$, δ)	**Quantitative Data ($\mu_1-\mu_2$, δ)**
Use *z/t* procedures if	Data are independent SRSs from populations of interest (H_0:$\pi_1-\pi_2=0$) or randomized experiment (H_0: $\delta=0$), and at least 5 successes and failures	Data are independent SRSs from populations of interest (H_0:$\mu_1-\mu_2$=hypothesized difference) or randomized experiment (H_0: δ=hypothesized effect), and either both populations distributions are normal or sample sizes ≥ 20 (robust)
Test statistic	$z_0 = \dfrac{\hat{p}_1 - \hat{p}_2}{\sqrt{\hat{p}(1-\hat{p})\left(\frac{1}{n_1} + \frac{1}{n_2}\right)}}$ where $\hat{p} = (x_1+x_2)/(n_1+n_2)$	$t_0 = \dfrac{(\bar{x}_1 - \bar{x}_2) - hypothesized\ difference}{\sqrt{s_1^2/n_1 + s_2^2/n_2}}$
p-value	From standard normal distribution	From *t* distribution (conservative df=*min*(n_1-1, n_2-1)
Confidence interval	$(\hat{p}_1 - \hat{p}_2) \pm z^* \sqrt{\dfrac{\hat{p}_1(1-\hat{p}_1)}{n_1} + \dfrac{\hat{p}_2(1-\hat{p}_2)}{n_2}}$	$(\bar{x}_1 - \bar{x}_2) \pm t^* \sqrt{s_1^2/n_1 + s_2^2/n_2}$
If *z/t*-procedures are not appropriate consider	Hypergeometric distribution	Bootstrapping or randomization
Also consider	$z_0 = \dfrac{(sample\ log\ odds\ ratio)}{\sqrt{\dfrac{1}{n_{11}} + \dfrac{1}{n_{12}} + \dfrac{1}{n_{21}} + \dfrac{1}{n_{22}}}}$ $sample\ log\ odds\ ratio \pm$ $z^* \sqrt{\dfrac{1}{n_{11}} + \dfrac{1}{n_{12}} + \dfrac{1}{n_{21}} + \dfrac{1}{n_{22}}}$	Bootstrapping/randomization for other statistics

TECHNOLOGY SUMMARY

- In this chapter you used Minitab to
    - Carry out two-sample *z*-procedures for the difference in proportions (population proportions and treatment effect)
    - Carry out two-sample *t*-procedures for the difference in means (population means and treatment effect)

You can also use the "Test of Significance Calculator" applet for the *z* and *t* procedures

CHAPTER 6: COMPARING SEVERAL POPULATIONS, EXPLORING RELATIONSHIPS

The idea of comparing two groups has been a recurring theme throughout this course. In the previous chapters, you have been limited to exploring two groups at a time. You saw that often the same analysis techniques apply whether the data have been collected as independent random samples or from a randomized experiment, although this data collection distinction strongly influences the scope of conclusions that you can draw from the study. You will see a similar pattern in this chapter as you extend your analyses to exploring two or more groups. In particular, you will study a procedure for comparing a categorical response variable across several groups and a procedure for comparing a quantitative response variable across several groups. You will also study the important notion of association between variables, first with categorical variables and then for studies in which both variables are quantitative. In this latter case, you will also learn a new set of numerical and graphical summaries for describing these relationships.

SECTION 6-1: TWO CATEGORICAL VARIABLES

In previous chapters you explored whether two population proportions were equal and whether two categorical variables were independent. Those methods were limited to *binary* variables. In this section, you will expand on your earlier techniques to allow for more than two categories in each variable.

Investigation 6-1: Dr. Spock's Trial

The well-known pediatrician and child development author Dr. Benjamin Spock was also an anti-Vietnam War activist. In 1968 he was put on trial and convicted on charges of conspiracy to violate the Selective Service Act (encouraging young men to avoid the draft). The case was tried by Judge Ford in Boston's Federal courthouse. A peculiar aspect of this case was that his jury contained no women. A lawyer writing about the case that same year in the Chicago Law Review said, "Of all defendants at such trials, Dr. Spock, who had given wise and welcome advice on child-bearing to millions of mothers, would have liked women on his jury" (Ziesel, 1969). The opinion polls also showed that women were generally more opposed to the Vietnam War than men.

In the Boston District Court, jurors are selected in three stages. The Clerk of the Court is supposed to select 300 names at random from the City Directory and put a slip with each of these names into a box. The City Directory is renewed annually by a census of households visited by the police, and it lists all adult individuals in the Boston area. In Dr. Spock's trial, this sample included only 102 women, even though 53% of the eligible jurors in the district were female. At the next stage, the judge selects 30 or more names from those in the box that will constitute the "venire." Judge Ford chose 100 potential jurors out of these 300 people. His choices included only 9 women. Finally, 12 actual jurors were selected after interrogation by both the prosecutor and the defense counsel. Only one potential female juror came before the court and she was dismissed by the prosecution.

In filing his appeal, Spock's lawyers argued that Judge Ford had a history of venires in which women were systematically underrepresented. They compared the gender breakdown of this judge's venires with the venires of six other judges in the same Boston court from a recent sample of court cases.

Records revealed:

	Judge 1	Judge 2	Judge 3	Judge 4	Judge 5	Judge 6	Judge 7	Total
Women on jury list	119	197	118	77	30	149	86	776
Men on jury list	235	533	287	149	81	403	511	2199
Total	354	730	405	226	111	552	597	2975

(a) Calculate the proportion of women on the jury list for each judge. Also create a segmented bar graph to compare these distributions. How do the judges compare?

(b) Let π_i = probability of judge i selecting a female for the jury list. State a null and alternative hypothesis for testing whether these data provide reason to doubt that the proportion of women on jury lists is the same for all seven judges.

Note: Your null hypothesis only states that the proportions are equal, we are not specifying a particular value for this common population proportion.

If the null hypothesis is true, then the long-run probability of a woman on the jury panel equals the same value for all 7 judges.

(c) Suggest an estimate for this common overall probability of a female juror.

(d) How many jurors did Judge 1 see? Suppose the long-run proportion of women in his juries was .261, how many of these jurors would you expect to be female? How many male?

(e) How many of the jurors that Judge 2 saw would you expect to be female if his long-run proportion was also .261? How many male?

(f) Continue to calculate the *expected number* of women and the expected number of men for each judge based on the total number of jurors in each column, assuming that $\pi = .261$ for each judge, entering the results below the *observed counts* in the table below. [*Hint*: The expected counts need not be integers.]

	Judge 1	Judge 2	Judge 3	Judge 4	Judge 5	Judge 6	Judge 7
Women on jury list	119	197	118	77	30	149	86
Men on jury list	235	533	287	149	81	403	511
Total	354	730	405	226	111	552	597

(g) Are the observed counts equal to the expected counts in each cell of the above table? Is it possible that the long-run probability of a female jury panel member is the same for each judge and we see the above differences just by chance?

(h) Suggest a way for measuring the overall deviation between the observed counts and the expected counts.

A common test statistic used to compare the observed and expected counts in a two-way table is the test statistic:

$$X^2 = \sum_{i=1}^{r} \sum_{j=1}^{c} \frac{\left(observed_{ij} - expected_{ij}\right)^2}{expected_{ij}}$$

where r = number of rows and c = number of columns. This test statistic looks at the discrepancy between the observed and expected counts, with any deviation giving a positive contribution to this sum, where each term is "standardized" by the expected size of the counts in that cell.

(i) Calculate this test statistic for the above two-way table. [Specify the value of each of the 14 terms you are summing together.]

(j) What types of test statistic values (large, small, positive, negative) constitute evidence against the null hypothesis of equal population proportions? Explain.

In order to approximate the p-value of this test, we need to examine how the test statistic behaves under the null hypothesis of equal population proportions. We will once again first explore this distribution by simulating a large number of samples where the probability of success is the same for each judge.

Open an empty Minitab worksheet and name columns C2–C8 "Judge1", "Judge2" up through "Judge7" (without the quotation marks). The macro `spockSim2.mtb` randomly generates the male/female breakdown for each judge's panels by sampling from a binomial distribution with n_i equal to judge i's sample size and assuming $\pi = .261$ for the probability of a juror being female for each judge. The resulting row totals for each gender are stored in C9. Then the expected counts are computed in rows 4–5 and the "observed" counts are compared to the excepted counts. Column 10 contains the 14 terms of the Chi-Square sum and the resulting Chi-Square statistic for each sample are stored in C11.

(k) Initialize `k1` and run the macro 500 times. Produce numerical and graphical summaries of the empirical sampling distribution of these test statistic values in C11. Describe the behavior of this distribution.

(l) What proportion of observations in this distribution is larger than the test statistic value you computed in (i)? Is this convincing evidence that the difference between observed and expected counts that you observed is larger than we would expect by chance? Explain.

Rather than rely on simulation to produce (approximate) p-values for this test, we can use a probability model to approximate the sampling distribution of the test statistic.

(m) Create a probability plot to see if the distribution in C11 is approximately normal. What is your assessment?

(n) Double click on the blue guidelines in the probability plot. Then in the Edit Distribution Fit window, select the Gamma distribution under the Options tab. What is your assessment of the fit of this model to the simulated test statistic values? What are the values of the Gamma distribution shape and scale parameters estimated by Minitab?

Probability Fact: A Gamma Distribution with parameters α and $\beta=2$ is also known as a *Chi-Square distribution* with parameter ν (which equals 2α). The parameter ν is referred to as the degrees of freedom of the Chi-Square distribution. The Chi-Square distribution is skewed to the right and provides a reasonable model to the above test statistic for large sample sizes. We will consider the Chi-Square distribution model appropriate if all of the expected counts are at least 1 and if at least 20% of the expected counts are at least 5.

When comparing several population proportions, the Chi-Square degrees of freedom is equal to the number of explanatory variable categories minus 1, $c - 1$. This makes sense—once we specify the number of observations in $c - 1$ of the categories, the last category is forced to assume the value that allows the observed counts to sum to the sample size.

(o) For the distribution fit, double click on the blue curve and then change the shape parameter to 3 and the scale parameter to 2. Is this still a reasonable model of the distribution in C11?

(p) For large sample sizes, we will use the Chi-Square probability distribution to approximate the p-value. Choose Calc > Probability Distribution > Chi-Square from the Minitab menu bar. Specify the cumulative probability with degrees of freedom 6. Specify 62.68 as the input constant and click OK. *Hint*: Remember to subtract the reported value from one to determine the probability *above* 62.68.

How does this p-value compare to the empirical p-value you determined in (l)?

Discussion: If the null hypothesis is rejected, the conclusion we draw is that at least one of the population proportions differs from the rest, but we don't have much information about which one. It could be one explanatory variable group is behaving much differently than the rest or they could all be different. One way to gain more information about the nature of the differences between the π_i values is to compare the components of the Chi-Square statistic sum.

(q) Return to the sum you calculated in (i). Which cell comparison(s) provide the largest (standardized) discrepancy between the observed counts and the expected counts?

(r) For the cells identified in (q), which is larger, the observed counts or the expected counts? Explain the implications of this comparison.

(s) Which judge do you believe tried Dr. Spock's case? Explain.

Study Conclusions: One judge clearly stood out compared to the others in these sample data. If we consider these results as representative of the overall jury selection process, the very small p-value indicates that if in fact the judges' selections of jurors were independent random processes with the same probability of selecting a woman, then it would be almost impossible to observe sample proportions differing by this much by chance alone. Thus, the sample data provide strong evidence that the population proportions are not the same among all seven judges. The largest contributions to the X^2 test statistic, by far, come from judge 7, who has many more men than would be expected and many fewer women than would be expected on his jury lists. This was indeed the judge assigned to Dr. Spock's case. In fact, there are two issues with this judge, the sampling from the city directory led to a far smaller percentage of women (29%) than the city population (53%) across all the judges and then the proportion of women selected by Judge Ford dipped even lower to around 15% women. By the way, the Court of Appeals reversed Spock's conviction on other grounds without reaching the jury selection issue. While this p-value is hypothetical in nature (there was not a true random mechanism generating these data), it still provides a measure of how surprising these results are.

Technical Conditions: The Chi-Square distribution is an approximation to the sampling distribution of the Chi-Square test statistic when the data arise from independent binomial random variables. This approximation is considered valid as long as the average expected cell count is at least 5 and all of the individual expected cell counts are at least one. Notice we are discussing the *expected* cell counts here, not the individual cell counts. The data also need to have been collected from independent random samples or a randomized comparative experiment.

Summary of Chi-Square Test of Homogeneity of Proportions

Numerical and graphical summaries: conditional proportions and segmented bar graphs

If the data are independent random samples from I different populations or processes, then the hypotheses

$H_0: \pi_1 = \ldots = \pi_I$ where π_i is the probability of success in population i.

H_a: at least one π differs from the rest

can be tested using the Chi-Square test statistic:

$$X^2 = \sum_{i=1}^{2} \sum_{j=1}^{I} \frac{\left(observed_{ij} - expected_{ij}\right)^2}{expected_{ij}}.$$

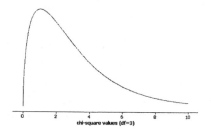

chi-square values (df=3)

The (upper-tail) p-value is calculated from the Chi-Square distribution with $(I-1)$ degrees of freedom.

In Minitab:

- If the two-way table has been entered into I columns, choose Stat > Tables > Chi-Square Test (Table in Worksheet) and specify the columns containing the table. Click OK.

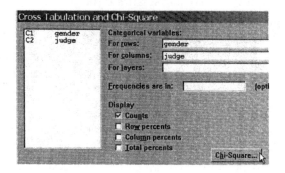

C1-T	C2	C3	C4	C5	C6	C7	C8
	Judge 1	Judge 2	Judge 3	Judge 4	Judge 5	Judge 6	Judge 7
women	119	197	118	77	30	149	86
men	235	533	287	149	81	403	511

- If the raw data for each categorical variable are contained in two different columns, choose Stat > Tables > Cross Tabulation and Chi-Square. Click the Chi-Square button and select "Chi-Square Analysis" to be displayed. Click OK twice.

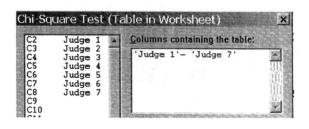

If we reject the null hypothesis, we only conclude that at least one π_i differs from the others. We can gain insight as to which proportion(s) are most different by comparing the relative contributions to the Chi-Square sum of each cell.

TYPE I ERRORS

One question that you may be asking is why not just use the two-sample z procedures to compare pairs of proportions? We could do this for two groups at a time, e.g., comparing Judge 1 to Judge 2 and then Judge 3 to Judge 5 and so on.

(t) How many such two-group comparisons are there among these 7 judges?

(u) If the level of significance is set to .05, what is the probability of a Type I Error for any one of these comparisons?

(v) What is the probability of *at least one* Type I Error among these 21 comparisons?

The advantage of a Chi-Square Test is that it provides an overall p-value for all the comparisons at once, which controls the overall probability of a Type I Error. If the p-value is not significant, we usually do not check all of the individual comparisons. If the p-value is significant, then we can do more formal follow-up analyses to see where the difference(s) are arising. If we run many tests on the same data set, we are always concerned with an inflated overall Type I Error rate.

Investigation 6-2: Near-Sightedness and Night Lights (cont.)

Recall the study published by Quinn, Shin, Maguire and Stone (1999) that examined the relationship between the type of lighting children were exposed to and subsequent eye refraction. Below is the two-way table and segmented bar graph that you explored:

	Darkness	Night light	Room light	Total
Hyperopia	40	39	12	91
Emmetropia	114	115	22	251
Myopia	18	78	41	137
	172	232	75	479

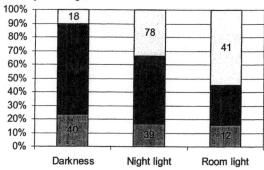

(a) What proportion of subjects in this study was classified as having Hyperopia (far-sightedness)? With Emmetropia (normal eyesight)? With Myopia (near-sightedness)?

(b) If the children's current eye conditions were not related to the level of lighting they were exposed to, what proportion of children in the darkness condition would have Hyperopia? Emmetropia? Myopia?

(c) What would the proportion breakdown of eye conditions look like in the Night light group and in the Room light group if there were no association between eye condition and lighting level?

(d) Use your answers in (b) and (c) to calculate the *expected number* of children in each cell of the table, assuming that there is no relationship between lighting and eye refraction. [*Hint*: For example, multiply the number of children in the darkness group by the proportion that would have hyperopia in that group if there were no relationship.] Record your answers in the table below next to the observed counts.

Expected counts:

	Darkness	Night light	Room light	Total
Hyperopia	(40)	(39)	(12)	91
Emmetropia	(114)	(115)	(22)	251
Myopia	(18)	(78)	(41)	137
	172	232	75	479

If we assume there is no relationship between the explanatory variable and the response variable, we can calculate the *expected count* for each cell by calculating

$$\text{Expected count} = \frac{\text{Row total} \times \text{Column total}}{\text{Table total}}$$

This produces the same distribution of conditional proportions in each explanatory variable group. This formula generalizes the approach you used in Investigation 6-1.

(e) Are the observed counts equal to the expected counts in each cell of the above table? Is it possible that there is no relationship between the response and explanatory variable but we might observe differences this big just by chance?

(f) Calculate a Chi-Square test statistic to measure the discrepancy between the observed and expected counts for this table

$$X^2 = \sum_{i=1}^{r} \sum_{j=1}^{c} \frac{\left(observed_{ij} - expected_{ij}\right)^2}{expected_{ij}}$$

where r = number of rows and c = number of columns. Using Minitab,

- Enter the two-way table into C1 – C3 (not including row labels or totals)
- Choose Stat > Tables > Chi-Square Test (Table in Worksheet)
- Select columns C1 – C3
- Click OK

What does Minitab report for the Chi-Square value, degrees of freedom and p-value? [You can also verify the expected counts that you computed in part (d).]

The degrees of freedom for the Chi-Square statistic with r rows and c columns is
$$df = (r-1)(c-1)$$

(g) Minitab also reports the Chi-Square contributions for each cell. Which cell(s) contribute the most to the overall Chi-Square sum? Compare the observed counts to the expected counts. What do these comparisons reveal about the nature of the relationship between children's eye condition and level of lighting?

Study Conclusions: We saw in chapter 1 that the incidence of myopia appeared to increase as the level of lighting used increased. Since table entries are large, we can apply the Chi-Square test to these data. The p-value of this Chi-Square test is essentially zero, which says that if there were no association between eye condition and lighting, then it's virtually impossible for chance alone to produce a table in which the conditional distributions would differ by as much as they did in the actual study. Thus, the sample data provide overwhelming evidence that there is indeed an association between eye condition and lighting. A closer analysis of the table and the Chi-Square calculation reveals that there are many fewer children suffering from myopia than would be expected in the "darkness" group and many more children suffering from myopia in the "room light" group. But remember the main lesson of this study from chapter 1—we cannot draw a cause-and-effect conclusion between lighting and eye condition because this is an observational study. Several confounding variables could explain the observed association. For example perhaps myopic children tend to have myopic parents who prefer to leave a light on because of their own eye condition, while also passing this genetic predisposition on to their children. We also have to be careful in generalizing from this sample data to a larger population since they were making voluntary eye doctor visits and were not selected at random.

Technical Conditions: The sample size conditions as checked by the expected cell counts for this Chi-Square procedure is the same as in Investigation 6-1. We also need the data to be a simple random sample from the population or process of interest, with the observational units cross-classified by the two categorical variables.

Summary of Chi-Square Test of Association

Numerical and graphical summaries: conditional proportions and segmented bar graphs

If the data are a random sample from a larger population and you have classified each observational unit according to two categorical variables, then the hypotheses:

H_0: no association between variable 1 and variable 2

vs. H_a: there is an association between variable 1 and variable 2

can be tested using the Chi-Square test statistic

$$X^2 = \sum_{i=1}^{r} \sum_{j=1}^{c} \frac{\left(observed_{ij} - expected_{ij}\right)^2}{expected_{ij}} .$$

The (upper-tail) p-value is calculated from the Chi-Square distribution with $(r-1)(c-1)$ degrees of freedom.

In Minitab:

- If the $r \times c$ two-way table has been entered, choose Stat > Tables > Chi-Square Test and specify the columns containing the table. Click OK.

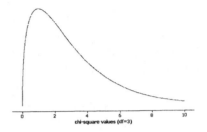

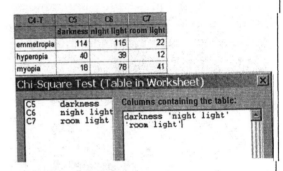

- If the raw data for each categorical variable are contained in two different columns, choose Stat > Tables > Cross Tabulation and Chi-Square. Click the Chi-Square button and select "Chi-Square Analysis" to be displayed. Click OK twice.

Investigation 6-3: Newspaper Credibility Decline (cont.)

Recall the newspaper believability ratings discussed in Investigation 5-1. In that activity, you looked at the binary variable "whether or not the rating was a 3 or a 4." In this activity you will explore how to compare the entire distribution of responses between the two years.

When asked about "The daily newspaper you are most familiar with," the percentage distribution of the 1002 responses in May 2002 was:

Believe all or almost all – 4	3	2	Believes almost nothing – 1	Can't Rate
20%	39%	25%	9%	7%

A similar study conducted in May 1998 yielded the following results (981 responses).

Believe – 4	3	2	Cannot Believe – 1	Can't Rate
27%	36%	24%	7%	6%

(a) Create a two-way table of counts to analyze the association between time and how people rate the believability of the daily newspaper they are most familiar with, *among those who felt they could rate their daily newspaper.* In other words, eliminate the "can't rate" category from consideration. Also remember to use the "explanatory variable" as the column variable and that the observed counts do need to be integers.

Discussion: If we have independent random samples from several populations as in Investigation 6-1, but the response variable has more than 2 categories ($r>2$), we can again extend the Chi-Square procedure. The expected counts and test statistic will be calculated exactly the same way as in Investigation 6-2. The Chi-Square distribution with $(r-1)(c-1)$ degrees of freedom will be valid with the same technical condition on sample size and the null hypothesis is that the distributions of the response variable are the same in all I populations. The alternative is that the I distributions are not all the same.

You should notice that the mechanics are exactly the same whether you are testing homogeneity of proportions, comparing the distributions across several populations, or examining the association between two categorical variables. The difference lies in how the data were collected and therefore in how the conclusions should be stated. In fact, these procedures are sometimes all phrased in terms of the association between variables, as this is mathematically equivalent to having the same conditional distributions. Again, you need to consider the implications of whether or not the distribution of a variable was determined by the researchers.

(b) State the null and alternative hypotheses, check the technical conditions, and carry out a Chi-Square test to assess whether the sample data provide strong evidence that the distribution of responses differs between the two years. State your conclusion in context.

COMPARISON TO TWO-SAMPLE Z-TEST

If we collapse the above table to focus on the "largely believable" (3 or 4) and "not largely believable" (1 or 2) breakdown, we obtain:

	2002 sample	1998 sample	Total
Largely believable	591	618	1209
Not largely believable	341	304	645
Total	932	922	1854

(c) Carry out a Chi-Square test to decide if the difference in the population proportions giving a largely believable rating differed significantly between the two years.

(d) How do the test statistic from the two-proportion z-test (question (e), Investigation 5-2) and the Chi-Square test statistic compare? How does the Chi-Square p-value compare to the *two-sided* p-value for comparing the two proportions?

Discussion:

The Chi-Square procedure can be used to compare two or more population proportions. When there are only two populations, the procedure is equivalent to a two-sided z-test for proportions from chapter 5. The Chi-Statistic is equal to z^2 and the Chi-Square p-value is equal to the two-sided p-value for the two-sample z test. If the alternative hypothesis is two-sided, you can use either procedure. If the alternative hypothesis is one-sided, then you should carry out the two-sample z-test to obtain the one-sided p-value. If there are more than two populations, then you must use the Chi-Square procedure and will only assess whether or not at least one population proportion differs from the others.

For a 2×2 two-way table, another alternative is Fisher's Exact Test as you learned in chapter 1. It is always appropriate to carry out Fisher's Exact Test, though it may be a bit less convenient with huge sample sizes (which becomes less of an issue with each new computer chip). However, of all three of these procedures, the two-sample z-procedure is the only one that also enables us to calculate a confidence interval to estimate the magnitude of the difference in the two population proportions.

Practice:

6-1) Chi-Square Test Statistic Simulation of Randomization
Suppose we wanted to analyze the Near-Sighted study as a pseudo-experiment and wanted to explore the treatment effects of the lighting group. Open the worksheet `refraction.mtw`. Column 1 indicates the response variable group (hyperopia, emmetropia, and myopia) and column 2 indicates the explanatory variable group (darkness, night light, room light). The table of expected counts has been placed into C5–C7.
(a) To recreate the two-way table of observed counts, choose Stat > Tables > Cross Tabulation and enter C1 as the row variable and C2 as the column variable. Click the Chi-Square button and check the "Chi-Square analysis" box.
The macro `twoway.mtb` has been created to randomize the assignment of subjects to the lighting groups (assuming no relationship between eye condition and lighting). Then the macro creates the new two-way table (storing those results in columns C10–C12) and calculates the Chi-Square test statistic for each simulated table (storing the results in C20). Open this macro in a text editor to see the commands.
(b) Explain the commands in the `twoway.mtb` macro. What is each Minitab command accomplishing and how does this produce the final distribution of interest?
(c) Initialize the counter (`let k1=1`) and execute the macro 1000 times. Create a graphical display of the resulting empirical sampling distribution of Chi-Square values. Describe the pattern of the distribution.
(d) Does this distribution appear to be reasonably modeled by a Chi-Square distribution with df = 4? Explain.
(e) Compute an empirical p-value based on these simulated Chi-Square values and compare it to the p-value calculated above.
(f) Use Minitab to compute the p-value for each of the simulated Chi-Square values (you can pass the entire column into Calc > Probability Distributions > Chi-Square, choosing the cumulative probability with 4 degrees of freedom, storing the results in C21). Describe the shape, center, and spread of the distribution of these p-values. Explain why this distribution makes sense.

(g) Suppose you had been given a 4×4 two-way table instead of a 3×3 table. Comment on what changes you would need to make in the `twoway.mtb` macro in order to simulate the sampling distribution of the Chi-Square statistic for a table of this size. (You do not need to carry out this simulation.)

6-2) Income and Back Pain

A recent study (Center on an Aging Society, March 2003) examined the relationship between median annual earnings of adults and work limitations due to back pain. The following two-way table displays the results for the 18 – 44 years in the study.

	Difficulties at work due to back pain	Have not experienced difficulties at work
<5K	21	688
5K-10K	21	1047
10K-15K	19	1303
15K-20K	16	1822
20K-25K	10	1800
25K-35K	23	3436
35K-50K	17	3314
50K+	9	2585

(a) Produce numerical and graphical summaries to describe the relationship in this sample.
(b) Is there evidence of a significant association between annual earnings and whether or not 18–44 year-olds experience difficulties at work due to back pain? Explain.

6-3) News Believability

In February of 1993, NBC News admitted that it staged the explosion of a General Motors truck during a segment of the program "Dateline NBC" in November of 1992. The segment included crash footage that explosively showed how the gas tanks of certain old GM trucks could catch fire in a sideways collision.

In a nationwide poll of adults (*Times Mirror News Interest Index*) conducted in August 1989, 1507 respondents gave NBC news the following believability ratings.

Believe – 4	3	2	Cannot Believe – 1	Can't Rate
32%	47%	14%	2%	5%

The same poll conducted Feb. 12–27, 1993 saw 2001 respondents give the following results:

Believe – 4	3	2	Cannot Believe – 1	Can't Rate
31%	42%	18%	6%	3%

(a) Is the difference in believability ratings statistically significant between these two years?
(b) Is this convincing evidence that the General Motors explosion caused a decrease in the believability of NBC News?

SECTION 6-2: COMPARING SEVERAL POPULATION MEANS

In the last section, you learned the Chi-Square test for comparing proportions among two or more groups. One key observation was that there are advantages to an overall procedure that compares the proportions simultaneously and controls the overall Type I Error rate. When there are only two groups, this procedure was equivalent to the two-sample two-sided z-test for proportions. You will see a similar approach in this section, addressing the issue of comparing two or more population *means*. The technique you learn here will be used both for comparing population means based on independent random samples and also for assessing whether there is a treatment effect based on data from a randomized experiment.

Investigation 6-4: Handicap Discrimination

The U.S. Vocational Rehabilitation Act of 1973 prohibited discrimination against people with physical disabilities. The act defined a handicapped person as any individual who had a physical or mental impairment that limits the person's major life activities. In 1984, handicapped individuals in the labor force had an unemployment rate of 7% compared to 4.5% in the non-impaired labor force.

Researchers conducted a study in the 1980's that examined whether physical handicaps affect people's perceptions of employment qualifications (Cesare, Tannenbaum, and Dalessio, 1990). The researchers prepared videotaped job interviews, using the same actors and script each time. The only difference was that the job applicant appeared with different handicaps:
- No handicap
- Leg amputation
- Crutches
- Hearing impairment
- Wheelchair confinement

Seventy undergraduate students were randomly assigned to view one of the videotapes, and they were then asked to rate the candidate's qualifications on a $1 - 10$ scale. The research question is whether subjects tend to evaluate qualifications differently depending on the applicant's handicap.

(a) Identify the observational units, explanatory variable, and response variable in this study. Identify each variable as quantitative or categorical. Is this an observational study or an experiment? Explain.

(b) The mean applicant qualification scores for the five groups are amputee 4.429, crutches 5.921, hearing 4.050, none 4.900, and wheelchair 5.343. What additional pieces of information do you need in order to decide if these sample means are significantly different (more than you might expect by sampling variability)? That is, do you believe these samples all came from the same population?

(c) State a null and an alternative hypothesis to reflect the researchers' conjecture.

(d) Explain what a Type I Error and a Type II Error represent in this context.

(e) Consider the following sets of boxplots. What is the same and what is different about the distributions displayed by these boxplots? In which group do you believe the evidence will be stronger that at least one population mean differs from the others? Explain.

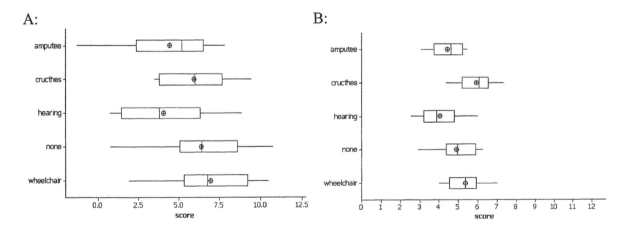

Discussion: In the above boxplots, the sample means are the same between the two cases (5.34, 5.92, 4.05, 4.43, 4.90) but the variability in the data is much larger for the distributions in graph A. This larger amount of "within sample" variation makes the difference in sample means look not as extreme, but when there is much less within sample variation as in graph B, the difference in the sample means appears more extreme and we have more evidence that the crutches scores and the hearing scores did not come from the same population with the same population mean. So our goal in comparing the sample means will be to decide whether the differences in the sample means are larger than what we would expect by random chance where the amount of within sample variation will give us an indication as how much we expect the values to vary "by chance."

(f) The actual data are stored in `HandicapEmployment.mtw`. Produce parallel boxplots and descriptive statistics of the qualification scores among the 5 groups. Record the descriptive statistics in the table below. Do these data suggest evidence of a difference among the population means? Write a paragraph supporting your statements.

	None	Amputee	Crutches	Hearing	Wheelchair
Sample size					
Sample mean					
Sample std. dev.					

As with the Chi-Square procedures, we need a test statistic that simultaneously measures the differences between several groups. As noted above, we will do this by comparing the variation in the sample means from the overall mean and see if that variation is much larger than the chance variation exhibited within the samples.

(g) What is the overall mean applicant qualification rating assigned by the 70 students?

(h) If we were to treat the 5 sample means as 5 observations, calculate the *variance* of these 5 values? [*Hint*: variance = (standard deviation)^2]

(i) Is it reasonable for each of these sample means to have the same relative contribution to our overall measure of variability between group means? Explain.

> Our measure of the *variability between groups* or "treatment effect" will be the *weighted variance* across the groups where the weights are the sample sizes.

(j) In this case, the sample sizes were equal, so multiply your answer to (h) by 14.

Between group variability =

In order to assess whether the sample group means vary enough to be considered statistically significant, we also need to consider the variability within each group. This will allow us to decide whether our deviations are larger than we would expect by chance. To measure the overall variability, we will calculate the *pooled variance* across the groups.

(k) Calculate the average variance across the 5 groups. [*Hint*: Square each group's standard deviation to obtain its variance, and then take the average of those 5 values.]

> When the sample sizes are not equal, the more general formula for this pooled variance, to weight by sample size, is:
>
> and provides an estimate of the overall "within group" variability which simplifies to the average of the variances when the sample sizes are equal.

(l) In calculating a pooled variance, you are implicitly assuming that the variability is the same across the groups. Why is that a reasonable belief here?

Our test statistic will look at the ratio of the variability between groups to the variability within groups.

(m) Calculate the ratio of your result from (j) divided by your result from (k).

(n) What is the smallest possible value that this ratio could ever assume? The largest?

(o) What types of values (large, small, positive, negative) will this ratio have when the null hypothesis is false; i.e., when the population means are not all equal?

(p) Explain how we could simulate a randomization test to approximate the distribution of this test statistic.

(q) The macro RandomHandicap.mtb takes these 70 rating scores and randomly assigns them to the five different handicap groups (so that we are assuming no genuine treatment effect). The macro then calculates the observed treatment effect (C9) and the pooled variance (C10) for each randomization. Run this macro 1000 times. Then in C11, compute the test statistic as the ratio between columns 9 and 10. Describe the shape, center, and spread of the distribution of these test statistics.

(r) Calculate an empirical p-value for the ratio you calculated in (m). What conclusion would you draw based on this p-value?

Probability Result: For large samples, this sampling distribution is well modeled by an F *distribution* with parameters $I - 1$ and $N - I$, the degrees of freedom of the numerator and denominator respectively.

Terminology Detour:

We will compare I group means, where each group has n_i observations. The overall sample size will be denoted by $N = \Sigma n_i$.

H_0: no treatment effect or H_0: $\mu_1 = \ldots = \mu_I$

H_a: there is a treatment effect or H_a: at least one μ_i differs from the rest

The between group variability will be measured by looking at the sum of the squared deviations of the group means to the overall mean, $\bar{x}$. Each group mean is weighted by the sample size of that group. We will refer to this quantity as the "sum of squares for treatment," SST.

$$SST = \sum_{i=1}^{I} n_i (\bar{x}_i - \bar{x})^2$$

We will then "average" these values by considering how many groups were involved. This quantity will be referred to as the "mean square for treatments."

$$MST = \frac{SST}{(I-1)}$$

Note, if we fix the overall mean, once we know $I - 1$ of the group means, the value of the Ith mean is determined. So the degrees of freedom of this quantity is $I - 1$.

The within group variability will be measured by the pooled variance. In general, each term will be weighted by the sample size of that group. We will again divide by an indication of the overall sample size across the groups. We will refer to this quantity as the "mean squares for error," MSE.

$$MSE = \frac{\sum_{i=1}^{I} (n_i - 1)s_i^2}{N - I} \quad \text{which has } N - I \text{ degrees of freedom.}$$

The test statistic is then the ratio of these "mean square" quantities:

$$F = \frac{MST}{MSE}$$

When the null hypothesis is true, this test statistic should be close to 1. So larger values of F provide evidence against the null hypothesis. The corresponding p-value comes from a probability distribution called the F *distribution* with $I - 1$ and $N - I$ degrees of freedom.

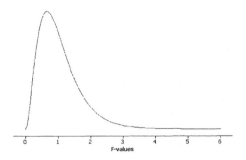

We will use this F distribution to approximate both the sampling distribution of this test statistic in repeated samples from the same population (H_0: $\mu_1 = \ldots = \mu_I$) and the randomization distribution for a randomized experiment (H_0: no treatment effect) as long as the technical conditions (see below) are met.

Since we are focusing on the variance of group means, this procedure is termed Analysis of Variance (ANOVA). With one explanatory variable, this is called *one-way ANOVA*.

(s) In Minitab, choose Stat > ANOVA > One-Way. Enter Score as the response variable and Handicap as the Factor variable. Click OK.

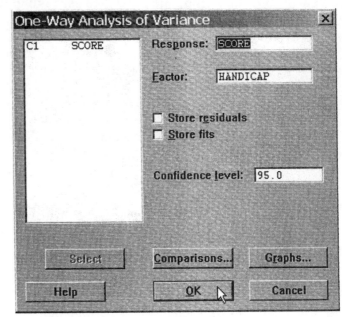

Note: The "(unstacked)" version would be used if each sample occurred in its own column.

Minitab reports the Mean Square values in the MS column, first for the treatment groups (the HANDICAP row) and then for the error term.

Verify that the F statistic is then the ratio of these two quantities.

You should also verify Minitab's calculations for the DF of the treatment row and of the error row.

Report the p-value:

There are several technical conditions that are required for this distribution to be well modeled by the F distribution:
- The distribution for each group comes from a normal population.
- The population standard deviation is the same for all the groups
- The observations are independent.

When using this F test to compare several population means, we will check each condition as follows:
- The normal probability plot for each sample is reasonably well-behaved
- The ratio of the largest standard deviation to the smallest standard deviation is at most 2
- The samples are independent random samples from each population.

For simplicity, we will apply the same checks for a randomized experiment as well (with the last condition being met if the treatments are randomly assigned). If the first two conditions are not met, then suitable transformations may be useful.

(t) Is there evidence of nonnormality in these samples? Is the ratio of the largest to smallest standard deviation less than 2?

(u) Write a paragraph summarizing your conclusions for this study.

Study Conclusions:

The samples look reasonably symmetric with similar standard deviations, so it is appropriate to apply the Analysis of Variance procedure. There is moderate evidence that the mean qualification ratings differ depending on the type of handicap (p-value = .030). Descriptively, the candidates with crutches appear to have higher ratings and the candidates with hearing impairments slightly lower ratings (other procedures could be used to follow-up to test the significance of these individual differences). This was a randomized experiment so we can attribute these differences to the handicap status but we must be cautious in thinking the students in this study are representative of a larger population, particularly, a population of employers who make hiring decisions.

Investigation 6-5: Restaurant Spending and Music

Recall the study described earlier (Practice 1-16) that examined the difference in spending at the Softley's restaurant when three different types of background music were played. Researchers wanted to know whether the background music would affect the amount spent at the meal. The following summary statistics were reported:

	Classical music	Pop music	No Music	
Mean	£24.13	£21.91	£21.70	
SD	£2.243	£2.627	£3.332	
	($n_1=120$)	($n_2=142$)	($n_3=131$)	N=393

(a) Based on these summary statistics, calculate the overall (weighted) mean amount spent and the pooled standard deviation. Check the consistency of your calculations with the sample data.

(b) State the null and alternative hypotheses corresponding to the researchers' conjecture.

(c) Based on the above summary statistics and part (a), calculate the F-statistic (by hand) and p-value (Calc > Probability Distributions > F).

(d) What additional information is necessary before you can continue to interpret this p-value?

Study Conclusions:
In this study, a larger average was spent while classical music was playing. With such a large test statistic and such a small p-value (F=31.48, p-value ≈ 0), we have strong evidence that the difference in the sample means did not arise by chance alone. However, we would need to assess the technical conditions before we could know if this inferential procedure is valid with these data. While the equal standard deviation assumption seems reasonable (3.332/2.243 < 2) we do not have the sample data to examine the shapes of the sample distributions. Still, the sample sizes are large and the *F* procedure is fairly robust to departures from the normality condition. Descriptively it appears that the average amount spent is larger when classical music is playing. This was a randomized experiment so we can draw a cause-and-effect conclusion from the music, but we would want to be very cautious in generalizing these results to other restaurants and even other times of the year.

EFFECTS OF SAMPLE SIZE

You will now explore the effects of such factors as the size of the difference in the population means, the overall population standard deviations, and the sample sizes on the test statistic and p-value.

Open the "ANOVA Simulation" applet at www.rossmanchance.com/applets/Anova/Anova.html. The sliders and text boxes should specify:
- 23 as each population mean
- 120, 142, and 131 as the sample sizes
- 3 as the population standard deviation.

(e) Click the Draw Samples button. A sample will be selected from each population. What F-statistic and p-value did you obtain?

(f) Click the Draw Samples button again. Did you obtain the same F-statistic and p-value? Why not?

(g) Click the Draw Samples button 10–20 more times, watching how the boxplots, F, and p-values change. Did you ever obtain a p-value below .05? Is this possible? Would it be surprising? Explain.

(h) Change the value of μ_1 to 24. Click the Draw Samples button 10–20 times. Do the resulting p-values tend to be larger or smaller than in (g)? Explain why this makes sense.

(i) Change each of the sample sizes to 20 and click Draw Samples 10–20 times. Do the resulting p-values tend to be larger or smaller than in (h)? Explain why this makes sense.

(j) Draw Samples until you have a p-value < .3. Now change the value of σ to 7. (Use the arrow to decrease the value in increments of .2 or drag the slider, watching how the p-value changes.) How does this affect the magnitude of the p-value? Explain why this relationship makes sense.

(k) How does the p-value generally change if you now continue to increase the value of μ_1? Explain why this relationship makes sense.

Discussion: If the null hypothesis is true, then the p-value should vary uniformly between 0 and 1. In this case, for example, the p-value will be less than .05 in 5% of all random samples, so 5% of samples would you be led to reject the null hypothesis even when it is true. However, when there truly is a difference in the population means, our ability to detect that difference based on sample data is affected by several factors:
- when the population means are further apart, the p-value is smaller
- when the within group variability is larger, the p-value is larger
- when the sample sizes are larger, and there is a true difference between the population means, then the p-value is smaller.
The p-value of any particular study is in essence random, so we need to remember the Type I and Type II Errors that we could be making. Type I error with Analysis of Variance indicates that we concluded that the population means differ when they really don't, and Type II error indicates that we failed to conclude that at least one population mean differs when the population means are in actuality not all equal.

Practice:

6-4) Dr. Spock Trial (cont.)
Another way to analyze the data for the trial of Dr. Spock is to look at the percentages of women on the different venires and determine whether the mean percentage of women is equal across the seven judges. Below are the percentages of women on the venires for a recent sample from each of the judges. These data are also in `SpockPers.mtw`.

Judge 1	Judge 2	Judge 3	Judge 4	Judge 5	Judge 6	Judge 7
16.8	27.0	21.0	24.3	17.7	16.5	6.4
30.8	28.9	23.4	29.7	19.7	20.7	8.7
33.6	32.0	27.5	21.5	23.5		13.3
40.5	32.7	27.5	27.9	26.4		13.6
48.9	35.5	30.5	34.8	26.7		15.0
45.6	31.9	40.2	29.5			17.7
32.5	29.8					18.6
33.8	31.9					23.1
33.8	36.2					15.2

(a) Explain what information we learn from analyzing the data this way that we did not see when we carried out the Chi-Square test on the overall proportion of women for each judge. Why might this information be useful?
(b) Produce numerical and graphical summaries to compare the percentages across the seven judges.
(c) Carry out an ANOVA to test if at least one judge has a different mean percentage. Did you state the null and alternative hypotheses in terms of population parameters or in terms of treatment effects?
(d) Comment on whether you believe the technical conditions for this procedure are met.

6-5) Handicapped Discrimination
Suppose the researchers had been most interested in comparing the results for those in a wheelchair to those with leg amputation.
(a) Carry out a two-sided two-sample pooled t test to assess whether there is a statistically significant difference in the average ratings assigned to these two groups.
(b) Carry out an analysis of variance to assess whether there is a statistically significant difference in the average ratings assigned to these two groups.
(c) How are the p-values from (a) and (b) related? How do you think the t test statistic and the F test statistic are related? Explain.
(d) Suggest a situation where the two-sample t procedure would be preferred and a situation where the ANOVA procedure would be preferred.

SECTION 6-3: RELATIONSHIPS BETWEEN QUANTITATIVE VARIABLES

In this section you will analyze data sets with two quantitative variables. The goal will be to describe the relationship between the variables. As always, you will start by learning some useful numerical and graphical techniques for summarizing the data. Then you will explore how to use the relationship to make predictions of one variable from the other. In the next section you will then move on to inferential techniques based on simulated sampling and randomization distributions as well as a mathematical model.

Investigation 6-6: House Prices

A group of students wanted to investigate which factors influence the price of a house (Koester, Davis, and Ross, 2003). They used www.househunt.com, limiting their search to single family homes in California. They collected a stratified sample by stratifying on three different regions in CA (northern, southern, and central), and then randomly selecting a sample from within each strata. They decided to focus on homes that were below 5000 square feet and sold for less than $1.5 million. The worksheet `housing.mtw` contains data for their final sample of 83 houses.

(a) Identify the observational units and the primary *response* variable of interest here. Also classify this variable as quantitative or categorical.

(b) Open the `housing.mtw` worksheet and produce numerical and graphical summaries of the price variable. Describe the distribution of housing prices in this sample. Does the shape of this distribution make sense? Explain.

(c) Based on the analysis in (b), if you were going to randomly select a house from this population of single-family homes in CA, what is your best prediction of the cost of this house? [*Hint*: Think about the *least squares* principle from chapter 2.]

Part of the students' investigation was to see if the price of the house was related to the size of the house (as measured by the total square footage of the house), the explanatory variable.

(d) Do you think there will be a relationship between these variables? Do you think larger homes will tend to be more or less expensive than smaller homes?

We will need a new graphical summary to visually explore the relationship between two quantitative variables, the *scatterplot*.

(e) To create a scatterplot in Minitab, choose Graph > Scatterplot, with the Simple option. Double click on Price to enter this as the Y variable and then double click on Sq. Ft. to enter this as the X variable. Click OK. [Or type `MTB> plot c3*c2` .] Describe the pattern displayed in this scatterplot. Does this pattern confirm your expectation in (d)?

Discussion: *Scatterplots* are useful for displaying the relationship between two quantitative variables. If one variable has been defined as the response variable and the other as the explanatory variable, we will put the response variable on the vertical axis and the explanatory variable along the horizontal axis.

In describing scatterplots you will describe the overall pattern between the two variables focusing primarily on three things:
- *Direction*: Is there a *positive association* (small values of *y* tend to occur at small values of *x* and large values of *y* tend to occur with larger values of *x*) or a *negative association* (small values of *y* tend to occur at larger values of *x* and vice versa)?
- *Strength*: How closely are the observations following the observed pattern?
- *Linearity*: Is the overall pattern in the scatterplot linear?

In the above scatterplot, there does appear to be an overall linear relationship with a positive slope indicating a positive association. The relationship does not appear to be all that strong as there are many houses that cost much more than we would expect based on their square footage.

Investigation 6-7: Drive for Show, Putt for Dough

Some have cited "Drive for show, putt for dough" as the oldest cliché in golf. The message is that the best way to improve one's scoring average in golf is to focus on improving putting, as opposed to, say, distance off the initial drive, even though the latter usually garners more ooh's and aah's. To see if this philosophy has merit, we need to examine whether there is a relationship between putting ability and overall scoring, and whether that relationship is stronger than the relationship between scoring average and driving distance. The file golfers.mtw contains the 2004 statistics (through the Honda Classic on March 20) on the top 80 PGA golfers, downloaded from http://www.pgatour.com/stats/index on March 20, 2004. Three of the variables recorded include:

- *Scoring average*: A weighted average, which takes the stroke average of the field into account. It is computed by adding a player's total strokes to an adjustment, and dividing by the total rounds played. This average is subtracted from par to create an adjustment for each round. Keep in mind that in golf low scores, as measured by number of strokes, are better than high scores.
- *Driving distance*: Average number of yards per measured drive. These drives are measured on two holes per round, carefully selected to face in opposite directions to counteract the effects of wind. Drives are measured to the point where they come to rest, regardless of whether or not they hit the fairway.
- *Putting average*: On holes where the green is hit in regulation, the total number of putts is divided by the total holes played.

(a) Do you expect the relationship between scoring average and driving distance to be positive or negative? Explain.

(b) Do you expect the relationship between scoring average and putting average to be positive or negative? Explain.

(c) Open golfers.mtw and examine a scatterplot of average score (c2) vs. driving distance (c9) and average score vs. average putts (c10). Describe each scatterplot. Do the relationships confirm your expectations in (a) and (b)? Does one relationship appear to be stronger than the other? If so, which?

To further analyze these data, we need a way of measuring the strength of the association between the two variables. We will do this with the *correlation coefficient.*

The following are the above scatterplots with the $\bar{x}$ and $\bar{y}$ lines superimposed.

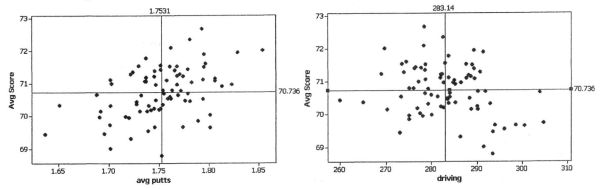

(d) For the average score vs. average putts scatterplot, in which quadrants are most of the points located? For the average score vs. driving scatterplot? Which scatterplot seems to have fewer points in the "non-aligned" quadrants?

Def: A numerical measure of the strength of a linear relationship is the *correlation coefficient, r.*

$$r = \frac{\sum\limits_{i=1}^{n}(x_i - \bar{x})(y_i - \bar{y})}{(n-1)s_x s_y}$$

where $\bar{x}$ and s_x are the mean and standard deviation of the explanatory variable, respectively, and $\bar{y}$ and s_y are the mean and standard deviation of the response variable.

Notice that when a point (x_i, y_i) is in quadrant I or III, the term $(x_i - \bar{x})(y_i - \bar{y})$ will be positive. When a point is in quadrant II or IV the term is negative. Where there is a positive association, most of the points are in quadrants I and III so the correlation coefficient is positive. The more observations that are in the "aligned" quadrants, the larger the value of the correlation coefficient.

(e) Let x represent the average number of putts and y represent the average number of strokes. Use the definition of the correlation coefficient above to determine the measurement units of the correlation coefficient r in terms of putts and strokes.

(f) If the correlation coefficient between two variables equals zero, what do you think the scatterplot will look like?

(g) Suppose we find the correlation coefficient of a variable with itself. Substitute x_i in for y_i (and so $\bar{x}$ for $\bar{y}$ and s_x for s_y) in the above equation. Simplify. What is the correlation coefficient equal to?

(h) Do you think the correlation coefficient will be a *resistant* measure of association? Explain.

(i) The following scatterplots display 7 relationships for these golfers. Rank these graphs in order from strongest negative correlation to strongest positive correlation.

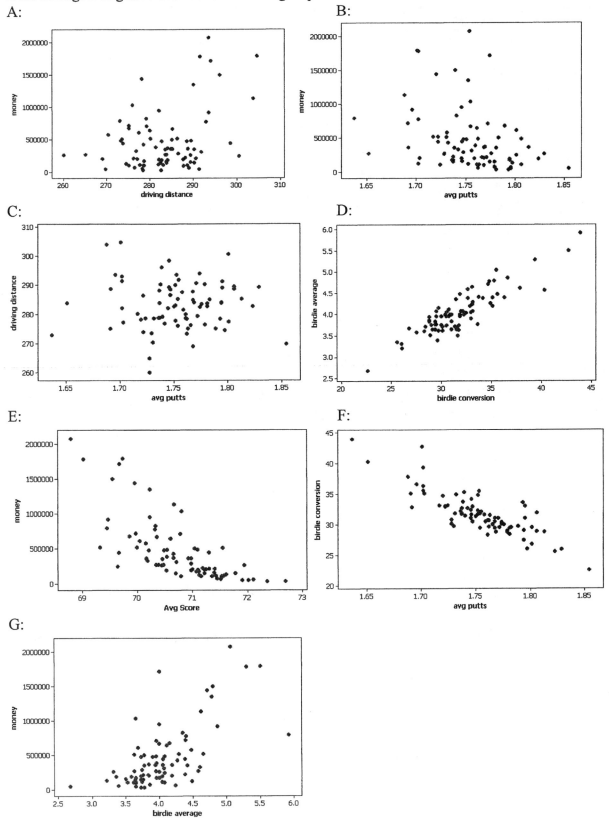

A:

B:

C:

D:

E:

F:

G:

(j) Use Minitab to determine the correlation coefficient for each of the above scatterplots by choosing Stat > Basic Statistics > Correlation and entering in the pair of variables. You can unselect the "Show p-values" box.

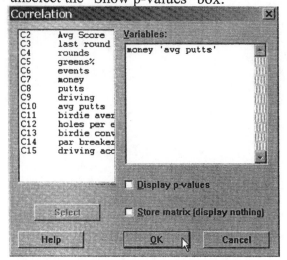

Alternatively, you can type:

MTB> corr c7 c10

Repeat for the pairs listed below.

Strongest negative   birdie conversion (C13) and average putts (C10)   _____

Medium negative   money (C7) and average score (C2)   _____

Weak negative   money (C7) and avg putts (C10)   _____

No association   driving distance (C9) and avg putts (C10)   _____

Weak positive   money (C7) and driving distance (C9)   _____

Medium positive   money (C7) and birdie average (C11)   _____

Strongest positive   birdie average (C11) and birdie conversion (C13)   _____

(k) Based on these observations and/or the formula, what is the smallest value (in absolute value) that r can assume? What is the largest value?

(l) If the association is negative, what values will r have? If the association is positive?

(m) What does a correlation coefficient equal to zero signify?

(n) What does a correlation coefficient close to 1 or −1 signify?

(o) Which has a stronger correlation, scoring average and driving distance or scoring average and average putts? Does this support the cliché? Explain.

Study Conclusions:

The correlation coefficient for scoring average and average putts indicates a moderately strong positive linear association ($r = .444$) whereas the correlation coefficient for scoring average and driving indicates a weaker negative association ($r = -.265$). This appears to support that putting performance is more strongly related to a PGA golfer's overall scoring average than the golfer's driving distance, as the cliché would suggest. We must keep in mind that these data are only for the first 2.5 months of the season (most golfers having only played around 6–8 events) and may not be representative of the scores and money earnings later in the year.

Discussion: The correlation coefficient, r, provides a measure of the strength and direction of the *linear* relationship between two variables. This is a unitless quantity that has the advantage that it will be invariant to changes in scale (if we started looking at money in British pounds instead of American dollars, our measure of the strength of the relationship will not change). A value of r close to zero indicates that the variables do not have a strong linear relationship. However, this does not preclude them from having a very strong but non-linear relationship. A scatterplot should be examined before interpreting the value of r. If the relationship is not linear, there are alternative measures of the strength of the association that can be used or the variables can be *transformed* and the transformed variables analyzed instead. If the relationship is linear, a correlation coefficient close to 1 or –1 indicates a very strong relationship. The values of r form a continuum: as the linear association becomes weaker, the value of r becomes closer to zero.

The correlation coefficient will always be a number between –1 and 1, inclusive. It will obtain the value of –1 or 1 if the points fall along a perfect line (with negative or positive slope).

Applet Exploration: Random Correlation Coefficients

Open the "Guess the Correlation" applet at www.rossmanchance.com/applets/GuessCorrelation.html

(a) Leave the number of points set to 15 and click New Sample. The applet will display a scatterplot. Guess the value of the correlation coefficient in this scatterplot and enter this guess into the "Correlation Guess" box and press Enter. Then click the "Show True Correlation" button. How close was your guess? Were you surprised by the actual value? In what way? How might you guess differently the next time?

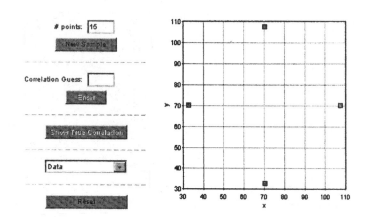

guess: .2   actual: .29

(b) Click the New Sample button, enter your guess for the value of the correlation coefficient for this scatterplot, and press Enter. Then press the button to show the true correlation. Describe how close you were and what adjustments you might make in your thinking.

guess: 0   actual: -.16

(c) Repeat this process for a total of 10 scatterplots (you do not need to record your results but you should try to learn from the reveal of the actual values as you make your guesses). Do you think your guessing ability improved by the last scatterplot? Explain.

look for the ries of the samples. try to imagine the line.

(d) Use the pull-down menu to show the "Guess vs. Actual" graph. Describe the behavior of the relationship between your guesses and the actual values of the correlation coefficients. Is there a strong correlation between your guesses and the actual values? Does this mean you are a good guesser? Explain.

fair guesser, sometimes over, sometimes under

(e) Use the pull-down menu to show the "Error vs. Actual" graph. Describe the behavior of the relationship between your errors and the actual values of the correlation coefficients. Were some correlation coefficients easier for you to guess than others? Use this graph to justify your answer.

less error cs corr. coeff → 1

more cs it went to zero

(f) Use the pull-down menu to show the "Error vs. Trial" graph. Describe the behavior of the relationship between your errors and the order in which you saw the graphs. Did your ability seem to improve over time? Use this graph to justify your answer.

as trial gets larger

error gets smaller, w/ some

exceptions, so I got better at guessing

(g) Suppose you guessed every value correctly, what would be the value of the correlation coefficient between your guesses and the actual correlations?

1

(h) Suppose each of your guesses was too high by .2 from the actual value of the correlation coefficient. What would be the value of the correlation coefficient between your guesses and the actual correlations?

1

(i) Does a correlation coefficient equal to 1 necessarily imply you are a good guesser? Explain.

it implys that you can get the

slope correct, but you can

be off by a constant.

Practice:

6-6) Wal-Mart Growth

The worksheet `walmart.mtw` contains data on the number of stores and the number of SuperCenters that Wal-Mart had in operation between 1989 and 2002.

(a) Use Minitab to produce a scatterplot of the number of stores (*y*-axis) versus time (*x*-axis). Describe the behavior of the association (in context). In particular, is the relationship strong or weak?

(b) Use Minitab to produce a scatterplot of the number of SuperCenters versus time. Describe the behavior of the association (in context). In particular, is the relationship strong or weak?

(c) Speculate as to the value of the correlation coefficient for each graph.

(d) Use Minitab to calculate each correlation coefficient. Would it be reasonable to conclude from the correlation coefficient that there is a moderate positive relationship between number of stores and time? Would it be reasonable to conclude from the correlation coefficient that there is a strong positive linear relationship between number of SuperCenters and time? Explain.

(e) Suppose in 2002 Wal-Mart had 1200 stores and 0 SuperCenters. Change these values in Minitab and recompute the correlation coefficients. Have they changed much? Explain. Is the correlation coefficient resistant to outliers?

6-7) Housing Prices (cont.)

(a) The following scatterplots look at the relationships between other variables in the real estate data set. How does the strength of the linear relationship between price and square footage compare to the strength of the relationships in the first 3 graphs?

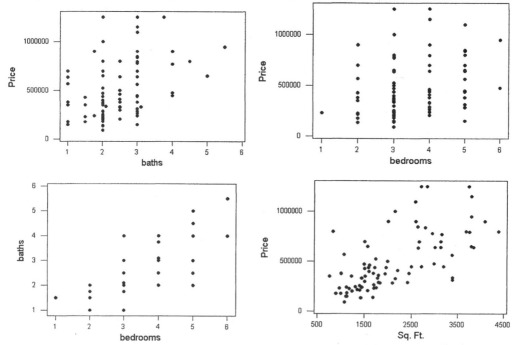

(b) The correlations for these four graphs are

.284, .394, .649, .760

Which correlation coefficient do you think corresponds to which graph? Explain your reasoning. [Note: Each graph has the same number of houses, but you may have multiple houses indicated by an individual dot.]

412

Investigation 6-8: Height and Foot Size

Criminal investigators often need to predict unobserved characteristics of individuals from observed characteristics. For example, if a footprint is left at the scene of a crime, how accurately can we estimate that person's height based on the length of the footprint?

(a) Identify the observational units, explanatory variable, and response variable in this study.

Observational units:

Explanatory variable:

Response variable:

Below are the heights (in inches) of 20 students in a statistics class:
74 66 77 67 56 65 64 70 62 67 66 64 69 73 74 70 65 72 71 63

(b) If you were trying to predict the height of a statistics student based on these observations, what value would you report?

(c) Using the method in (b), would you always predict a statistics student's height correctly?

Def: A *residual* is the difference between the predicted value and the observed value. If we let y_i represent the ith observed value and $\hat{y}_i$ represent the predicted or "fitted" value, then
$$\text{residual}_i = y_i - \hat{y}_i$$

(d) Calculate the residual for each of the above heights if we use the mean height, 67.75 inches, as the predicted value for each person. How often did we overestimate? How often did we underestimate?

74	66	77	67	56	65	64	70	62	67

66	64	69	73	74	70	65	72	71	63

(e) What is the relationship between the observed value and the fitted value if the residual is positive? What is the relationship between the observed value and the fitted value if the residual is negative?

(f) How might you measure the overall prediction error? [*Hint*: Recall that this is the same issue you faced with predicting house prices in Investigation 2-2 of chapter 2.]

Below is the scatterplot of the height (in inches) and foot length (in centimeters) for the sample of 20 statistics students.

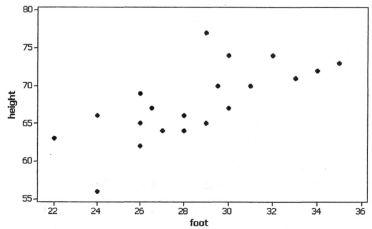

(g) Describe the association exhibited in this scatterplot. Does it behave as you would have expected? Explain.

(h) Sketch a line on the scatterplot that appears to summarize the overall linear relationship. Approximate the equation for your line.

In this bivariate setting, a residual is the vertical distance between the observed value and what we would predict using this line, residual$_i = y_i - \hat{y}_i$ where $\hat{y}_i = b_0 + b_1 x_i$.

(i) Use your line to predict the height of someone with a 28 cm foot length? What is the residual of your prediction?

(j) Where are the points with positive residual values located? Negative residual values?

(k) Did everyone in your class make the same prediction in (i)?

(l) Does your line provide a better fit than your neighbor's? Suggest a criterion for deciding which line "best" summarizes the relationship.

These data have been entered into Columns A and B of the Excel file `heightfoot.xls`. The scatterplot is displayed with a line at height $\bar{y} = 67.75$. The predicted heights ($\hat{y}_i$) based on this line are reported in Column C. In Column D, we examine the difference between the actual heights and the predicted heights, in absolute value $| y_i - \hat{y}_i |$. Column E squares each of these values, $(y_i - \hat{y}_i)^2$. The sum of these squared errors is reported in F10. The sum of the absolute errors, $\Sigma | y_i - \hat{y}_i |$, is reported in F9.

(m) Report the sum of absolute residuals (SA_{resid}) and sum of squared residuals (SS_{resid}) from this $\bar{y}$ line, using $\bar{y}$ as our prediction for every observation.

$SA_{resid}(\bar{y}) =$

$SS_{resid}(\bar{y}) =$

Our goal is to see if we can reduce these prediction errors (residuals) by using a line that takes the foot size of the individuals into account instead of only using the average height $\bar{y}$.

(n) In Column G, you can change the values of the slope and the intercept of the line displayed in the scatterplot. Change these values (a bit a time, note where the intercept is) until you think the resulting line does a reasonable job of summarizing the overall relationship between these two variables. Report the equation of your line, using variable names (instead of x and y) and indicate that the response variable is being estimated by placing a "hat" over the variable.

(o) Report the SA_{resid} and SS_{resid} for your line.

SA_{resid} (your line) =

SS_{resid} (your line) =

(p) Provide an interpretation in context for the two coefficients in your line.

Intercept =

Slope =

Discussion: A reasonable interpretation of the slope in this context would be the average change in height (inches) for each additional cm in foot length. If one person's foot length is one cm longer than another's, we predict this person to be this much (the slope coefficient) taller than the other person. Notice we are being careful to talk about the "average" change or the "predicted" change, as these are estimates based on the fitted line, not an exact mathematical relationship between height and foot length. Using this logic, the intercept here could be the predicted height for a person whose foot length is zero—not a very sensible statement in this context. In fact, the intercept will often be too far outside the range of x values for us to seriously consider its interpretation.

(q) Suggest a technique other than eyeballing and guess and check that would allow us to find the values of the slope and the intercept that minimize the SS_{resid}.

The Least Squares Regression Line

The least squares line $\hat{y} = b_0 + b_1 x$ is determined by finding the values of b_0 and b_1 that

minimize the sum of the squared residuals, $SS_{resid} = \Sigma(y_i - \hat{y}_i)^2 = \sum_{i=1}^{n}(y_i - b_o - b_1 x_i)^2$.

(r) Take the derivative with respect to b_0 of the expression on the right. Then take the derivative with respect to b_1. [*Hints*: Use the chain rule, and remember to treat the x_i's and y_i's as constants.]

(s) Set these (partial) derivatives equal to zero and solve simultaneously for the values of b_0 and b_1.
[*Hints*: Solve the first equation for b_0. Then solve the second equation for b_1 (substituting in the expression of b_0 using the summation notation).]

You should have found expressions like the following.

$$b_0 = \frac{\sum\limits_{i=1}^{n} y_i - b_1 \sum\limits_{i=1}^{n} x_i}{n} \quad \text{and} \quad b_1 = \frac{n \sum\limits_{i=1}^{n} x_i y_i - \sum\limits_{i=1}^{n} x_i \sum\limits_{i=1}^{n} y_i}{n \sum\limits_{i=1}^{n} x_i^2 - \left(\sum\limits_{i=1}^{n} x_i\right)^2}$$

With a little bit more algebra, you can show that the formulas for the least squares estimates of b_0 and b_1 simplify to:

$$b_0 = \bar{y} - b_1 \bar{x}$$
$$b_1 = r\, s_y/s_x$$

(t) For the above data set, $\bar{x} = 28.5$ cm, $s_x = 3.45$ cm, $\bar{y} = 67.75$ in and $s_y = 5.00$ in, with $r = .711$. Use these statistics to calculate the coefficients of the least-squares regression line.

(u) Specify these values in the Excel file. Report the resulting values of SA_{resid} and SS_{resid}. Are these smaller than the values you obtained by eyeball and guess-and-check?

SA_{resid} (least-squares):

SS_{resid} (least-squares):

One of the main purposes of finding a bivariate model for your data is to make predictions about one variable (the response variable) based on an observed value of the other variable (the explanatory variable). We want to find the model that minimizes some reasonable function of our prediction "error." The *least squares regression line* uses the values of the intercept b_0 and the slope b_1 that minimize the sum of the squared residuals: $\Sigma(y_i - \hat{y}_i)^2$. These coefficients do not necessarily minimize the sum of the absolute errors, but there are many theoretical results that apply for this least squares criterion that do not apply when using other criterion.

(v) Use the least squares regression line to predict the height of someone whose foot length is 28 cm. Will this prediction necessarily have a smaller residual than the one you made in (i)? Explain.

(w) Use the least squares regression line to predict the height of someone whose foot length is 44 cm. Explain why you should not be as comfortable making this prediction as the one in (v).

One way to assess the usefulness of this least-squares line is to measure the improvement in our predictions by using the least-squares line instead of the $\bar{y}$ line that assumes no knowledge about the explanatory variable.

(x) Report the SS_{resid} obtained in (m) when using only $\bar{y}$ to make our predictions: _____ Report the SS_{resid} obtained by the least squares regression line to make the predictions: _____

What is the *percentage change* in the SS_{resid} when we take the explanatory variable into account?

$$100\% \times [SS_{resid}(\bar{y}) - SS_{resid}(\text{least-squares}))/SS_{resid}(\bar{y})] =$$

Def: The above expression indicates the reduction in the prediction errors from using the least squares line instead of the $\bar{y}$ line. This is referred to as the *coefficient of determination*, denoted by r^2, and is interpreted as the percentage of the variability in the response variable that is explained by the least-squares regression on the explanatory variable. This provides us with a measure of how accurate our predictions will be and is most useful for comparing different models (e.g., different choices of explanatory variable).

Study Conclusions: There is a fairly strong positive linear association between the foot length of statistics students and their heights ($r = .711$). To predict heights from foot lengths, the least-squares regression line is $\widehat{height} = 38.3 + 1.03\,foot$. This indicates that if a foot length measurement is one centimeter longer than another, we will predict that person's height to be 1.03 inches taller as well. The slope tells us the average or predicted increase in height for each additional inch in foot length. This regression line has a coefficient of determination of 50.6%, indicating that 50.6% of the variability in heights is explained by this least squares regression line with foot length. So while the foot lengths are informative, they will not allow us to perfectly predict the heights of the students in this sample.

In Minitab

To use Minitab to calculate the least-squares line directly, choose Stat > Regression > Regression and specify the response variable in the first box and then the explanatory variable in the Predictors box.

To superimpose the regression line on the scatterplot, choose Stat > Regression > Fitted Line Plot.

Minitab reports additional output, but you should be able to find the least-squares regression equation and the value of r^2.

Click the Storage button and check the Residuals box to store them in their own column.

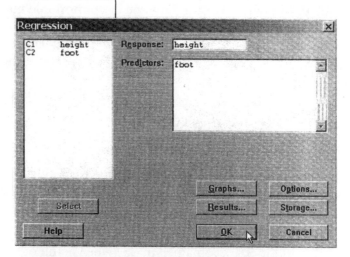

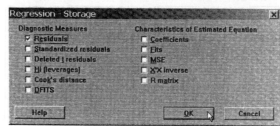

Investigation 6-9: Money Making Movies

Do "better" movies earn more money at the box office? *USA Today* (Wloszczyna and DeBarros, Feb. 25, 2004) investigated this by determining a rating score for movies released in 2003 based on a compilation of movie reviews published in 20 major newspapers and magazines for over 300 movies. The worksheet movies03.mtw contains these scores (C2) and how much money the movie made at the box office (C5), in millions of dollars. A high composite score indicates that most critics loved the movie, and a low score indicates that most critics panned the movie.

(a) Produce a scatterplot to determine if the critic scores appear to be related to the box office gross. Which did you treat as the explanatory variable and which as the response variable? Describe the relationship between the two variables as exhibited in the scatterplot.

(b) Identify the 2–3 points that you believe have the largest residuals. Use the "brush" feature to identify those movies by name. What does it mean for these movies to have such large residuals?

(c) Determine, report, and interpret the value of the correlation coefficient.

(d) Use Minitab to determine the least squares regression line for predicting the box office gross from the composite critics' score. Report the equation for the line and interpret the slope and the intercept for your equation in this context.

(e) Report the value of r^2 and provide an interpretation in this context.

When examining the relationship between two quantitative variables, we can also bring in information about a categorical variable.

(f) In Minitab, choose Graph > Scatterplot and choose the "With Groups" option. Enter the box office values and the ratings for the movies as before. In the Categorical variables box enter the "rating" variable (PG-13 etc) and then click the box next to "X–Y pairs from groups." The resulting scatterplot should have the movies coded by rating. Right click on the scatterplot and choose Add > Regression Fit. Click OK. Describe any patterns you see. In particular, does any rating category tend to have higher box office values than you would expect for the score they received from the critics? What about a rating category with lower box office revenues than expected? Explain.

(g) Repeat (f) using the movie genre (e.g., comedy).

(h) We can also superimpose the regression line on the scatterplot directly. Choose Stat > Regression > Fitted Line Plot and enter the "box office" revenue as the response and the "score" as the explanatory and click OK.

(i) In analyzing these movies, the researchers also looked at the data set after removing the 6 movies that earned more than $200 million. These data have been stored in C8–C12. Examine the scatterplot, correlation coefficient, and least-squares regression line for predicting the movie revenue from the critics' rating score for these data. Describe the effect of removing these observations on the analysis.

Def: An observation or set of observations is considered *influential* if removing the observation from the data set substantially changes the values of the correlation coefficient and/or the least squares regression equation. Typically, observations that have extreme explanatory variable outcomes (far below or far above $\bar{x}$) are potentially influential. To measure the influence of an observation, it is removed, and measures are calculated for how much the summary results change. It is *not* always the case that the points with the largest residuals are the most influential.

In this example you should have seen that removing those six movies actually makes the relationship look much weaker (r drops from .42 to about .3).

Study Conclusions: There does appear to be a weak relationship between the composite critics' scores and the amount of money the movie makes at the box office, with higher rated movies making more money. If the composite critics' score is 10 points higher, we predict the movie will make about 18.57 million more dollars. It is interesting to note that this regression line will tend to overestimate the amount of revenue for an R rated movie and underestimate the revenue of action movies. If the top 6 grossing movies of 2003 (Bruce Almighty, Finding Nemo, Pirates of the Caribbean, The Lord of the Rings III, The Matrix Reloaded, and X2: X-Men United) are removed, the relationship is not as strong, but still shows a weak positive linear association ($r = .3$).

Applet Exploration: Behavior of Regression Lines

Open the "Behavior of Regression Lines" applet (www.rossmanchance.com/applets/LRApplet.html). You will notice a scatterplot with 10 points and the least squares regression line.

(a) Does the regression line appear to do a reasonable job of summarizing the overall linear relationship in these data observations?

I suppose a " reasonable " job

(b) Click on the point located at about (3.22, 1.150), it should turn yellow. If we were to move this point to say (3.22, –3.0), predict whether, and if so how, the regression line will change.

Slope ↓

(c) Hold the mouse button down and move the mouse vertically in both directions to change the *y* value of the observation. The applet automatically recalculates the new regression line depending on the new location of the point. Is it possible to make the regression line have a negative slope? Does the regression line appear to be affected by the location of this point? Is the impact strong or weak? Does this match your prediction?

Can make n < 0   1 pt has

strong pull on line

(d) Click Reload in your browser and the point should be moved back to its original location. Now focus on the point located at (–.04, –1.02). If we move this point vertically, predict how the regression line will change. Do you think the change will be as dramatic as in (c)?

not as dramatic

(e) Repeat (c) using the point located at (–.04, –1.02). Does the regression line appear to be affected by the location of this point? Is the impact strong or weak (especially compared to the impact you witnessed in (c))?

changes the y - int of line

(f) Which point was more *influential* on the equation of the regression line (3.22, 1.150) or (–.04, –1.02). Suggest an explanation for why the point you identified is more influential, keeping in mind the "least-squares criterion."

Something is multiplied by x so

pt on end contributes more.

Practice:

6-8) Height and Foot Size (cont.)
(a) Report the value of the correlation coefficient and the coefficient of determination for these data. What symbols are used to refer to each value?
(b) If the value of the correlation coefficient is squared, how is this related to the value of the coefficient of determination?
(c) What would need to be true about the data set for the value of the coefficient of determination to be equal to 100%?

6-9) The 2000 Presidential Election
The 2000 U.S. Presidential election is infamous for the close outcome and the confusing "butterfly ballot" which some voters claimed led them to inadvertently select a candidate for whom they did not intend to vote (Pat Buchanan instead of Al Gore). In particular, Palm Beach County recorded an unusually large number of Buchanan votes. One way to examine evidence for this claim is to look at the number of votes given to the candidates in different counties.
(a) Open the 2000pres.mtw worksheet. Produce and interpret graphical and numerical summaries of the relationship between the number of votes for Buchanan in the different counties and the number of votes for Bush in these same counties.
(b) Use Minitab to determine the least-squares line for predicting the number of Buchanan votes from the number of Bush votes. Provide an interpretation of each coefficient in this context. Also report and interpret the value of r^2.
(c) Is there preliminary evidence that Buchanan received more votes than expected in Palm Beach County? Explain. Does Palm Beach County have the largest *residual* value?
(d) Remove Palm Beach County from the data set and recalculate the correlation coefficient, the least squares line and r^2. Do these measures appear to be *resistant*? Explain.
(e) Find the observation for Dade County. In what way is Dade County unusual in this data set?
(f) Remove Dade County from the data set and recalculate the correlation coefficient, the least squares line and r^2. Would you say that Dade County was *influential* in the calculations of these values? Explain.
(g) Now remove Sarasota County and recalculate the correlation coefficient, the least squares line and r^2. Would you say that Sarasota County was an influential observation?

6-10) Resistance
Which do you think will be most resistant to outliers, the regression line that minimizes the sum of squared errors or the regression line that minimizes the sum of the absolute errors? Explain.

SECTION 6-4: INFERENCE FOR REGRESSION

In the previous section you learned new techniques for describing the relationship between two quantitative variables, namely scatterplots and correlation coefficients, as well as a method for fitting a linear model to the data set. In this section, you will again make the transition to inference: what can we say about a larger population or about the statistical significance of the association based on the sample results? We will first develop a more general model that will allow us to describe the sampling distributions of the regression coefficients, and then we will discuss how to check the appropriateness of that model.

Investigation 6-10: Boys' Heights

Suppose we have data on the height (in centimeters) of boys for 3 different random samples at ages 2, 3, 4. The data in `hypoHt.mtw` are modeled after data from the Berkeley Guidance Study that monitored the height and weight of boys and girls born in Berkeley, California between January 1928 and June 1929.

(a) Which is the explanatory variable and which is the response variable?

Explanatory variable:

Response variable:

(b) Do you expect the 2-year-old boys to all have the same height? Do you expect the 3-year-old boys to all have the same height? Do you expect the mean height of the 3-year-old boys to be the same as the mean height of the 2-year-old boys? Explain.

Minitab reports the regression equation for this sample to be $\hat{height} = 75.2 + 6.47 age$, $r = .857$.
(c) Is it possible that in the population of Berkeley boys in the 1920s that there really is no linear relationship between age (2–4 years) and height but that we obtained a correlation coefficient and sample slope coefficient this large just by chance? Explain.

(d) Describe how we might investigate the likelihood of this occurring by chance.

(e) If there were no relationship between height and age in the population, what would this imply about the value of the population slope?

Terminology Detour:

To differentiate between the population regression line and the sample regression line, we will continue to use b_0 and b_1 to denote the sample statistics (or alternatively $\hat{\beta}_0$ and $\hat{\beta}_1$). The corresponding population parameters will be denoted by β_0 and β_1.

In order to determine whether our sample slope is significantly different from a conjectured value of the population slope, we need to assume a mathematical model.

To understand the most basic regression model, consider conditioning your data set on a particular value of x. For example, let's look at the distribution of heights for each age group.

(f) Open the hypoHt.mtw worksheet and create a dotplot (with groups) using the heights as the Graph variable and age as the Categorical variable. Compare the shape, center, and spread of the three distributions:

- How would you describe the shape of each distribution?

- Numerically, how do the means differ for the three distributions? Has the average height increased roughly the same amount each year?

- Is the variability in the heights roughly the same for each age group?

The most ***basic regression model*** specifies the following conditions:

- The relationship between the mean of the response variable and the explanatory variable is linear: $E(Y$ at each $x) = \beta_0 + \beta_1 x$
- The variability of the observations at each value of x is constant $SD(Y$ at each $x)$. We will refer to this constant value as σ. This means that the variability of the (conditional) response variable distributions does not depend on x.
- The distribution of the response variable at each value of x is normal.

To determine whether this mathematical model is appropriate we will check for three things: linearity, equal variance, and normality.

(g) Do these conditions appear to be met for the Berkeley boys' heights? Explain.

Discussion: In the above example, checking the conditions was straight forward since we could easily examine the distribution of the heights at the 3 different values of *x*. When we don't have many repeat values at each value of *x* we will need other ways to check these conditions. However, you should notice that the only thing that changes about the distribution of the response as we change *x* is the mean. So if we were to subtract the mean from each observation and pool these values together, then we should have one big distribution with mean 0, standard deviation equal to σ, and a normal shape.

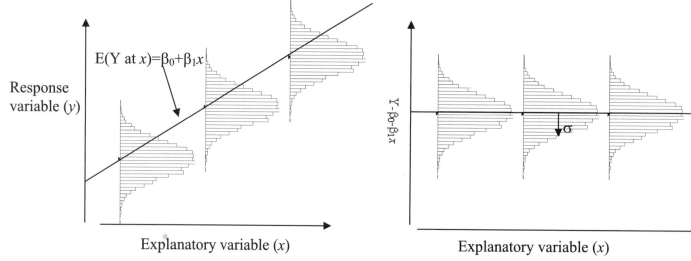

However we don't know β_0 and β_1, the population regression coefficients, but we have their least squares estimates. If we subtract the fitted values from each response value, these differences are simply the residuals from the least squares regression. So to check the technical conditions we will use *residual plots*.

- The linearity condition is considered met if a plot of the residuals vs. the explanatory variable does not show any patterns such as curvature.
- The equal variance condition is considered met if a plot of the residuals vs. the explanatory variable shows the same spread across the values of *x*.
- The normality condition is considered met if a histogram and normal probability plot of the residuals look normal.

- Another condition is that the observations are independent. We will usually appeal to the data being collected randomly or randomization being used. The main thing is to make sure there is not something like time dependence in the data.

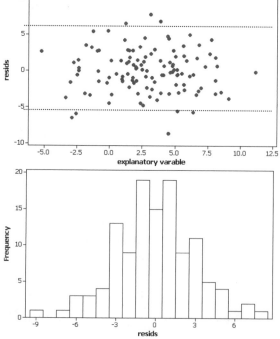

The above graphs show data that satisfy these conditions nicely. You will see some violations, and what we can do about them, in the next activity.

The above graphs show data that satisfy these conditions nicely. You will see some violations, and what we can do about them, in the next activity.

Investigation 6-11: Housing Prices (cont.)

Recall the California home prices randomly obtained by a student project group.

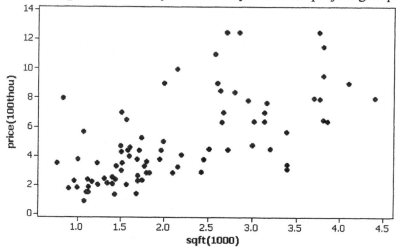

(a) Open the housing.mtw file and determine the least-squares regression line for these data. Also report the value of r^2 for these data.

(b) Based on the scatterplot, do these data appear to fit the basic linear regression model?

(c) Run the regression again, but this time click on the Storage button. In the Regression – Storage window, select the Residuals box.

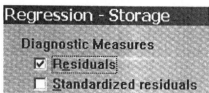

Minitab will automatically create a column with all of the residual values calculated (RESI1). Produce a histogram and normal probability plot of these residuals. Is it reasonable to consider the residuals as following the normal distribution? Explain.

(d) Produce a scatterplot of the residuals vs. the square footage.
- Does there appear to be curvature in this graph?

- Does the spread in the residuals appear to be roughly constant across the graph?

Discussion: It takes a while to become comfortable interpreting these residual plots, but the housing data do appear to have some problems. The residuals appear to cluster closer to zero for the smaller houses and to have more variability for the larger houses. This violates the condition of a constant standard deviation for each value of x. There is also a very, very slight hint of some curvature in this graph. The distribution of the residuals is clearly non-normal. These observations give us several reasons to not apply the basic regression model to these data.

> When the regression model conditions are violated, we can often *transform* the data in the hopes that the new variables do meet the conditions.

(e) Take the log base ten of both the prices and the sizes, storing the results in C10 and C11:

```
MTB> let c10 = logten('Price')
MTB> let c11 = logten('Sq. Ft.')
MTB> name c10 'logprice' c11 'logsqft'
```

Produce the scatterplot and find the regression equation to model the relationship between these two variables, storing the residuals. Examine the residual plots (histogram of residuals and residuals versus explanatory variable). Do these transformed variables appear to be more suitable for the basic regression model? Explain.

(f) Interpret the slope coefficient in the regression model using the logged variables.

(g) Use the least squares regression line to predict the cost of a 3,000 square foot house. [*Hints*: Substitute $\log_{10}(3000)$ into the right hand size to obtain a prediction for the $\log_{10}$price, and then raise 10 to this power to find the predicted price.]

Discussion: The log transformation has made improvements for both the normality assumption and the equal variance assumption. It is not uncommon for one transformation to "correct" more than one problem. We could continue to explore other transformations (e.g., square root and power transformations) until we find data that are more appropriate for the basic regression model. While these transformations are useful, we then have to be careful in how we interpret the regression coefficients and make predictions.

Investigation 6-12: Hypothetical Housing Prices

Using the transformed variables, the least-squares regression equation is

$\widehat{\log price} = 2.698 + .890 \log sqft$. Minitab also tells us the following summary statistics for each
variable: log(price)   mean=5.62   std dev = .260
                log(sqft)   mean=3.28   std dev = .192

(a) Is it possible that there is no relationship between log(price) and log(sq ft) in the population, but that we obtained a sample slope coefficient this large just by chance? Explain.

(b) If there was no relationship between log(price) and log(sqft) in the population, what would this imply about the value of the population slope coefficient β_1?

(c) State a null hypothesis and alternative hypothesis for testing whether the relationship between housing price and size is positive in this population (in symbols and in words).

Open the "Simulating Regression Lines" applet at
www.rossmanchance.com/iscat/applets/regcoeff/regcoeff.html
We will assume this is a population of homes, where
there is no relationship between house price and size.
To match the above context, we have:

- set the initial intercept to be equal to the mean of the log(price) variable. Since we are assuming $\beta_1=0$, this says that the population regression line is constant at $\bar{y}=5.62$
- set the x mean and x standard deviation to match the values given above for log(sqft)
- set the value of σ, representing the variability about the population regression line, to equal the standard deviation the log(price) variable.

So the population we are creating here matches the characteristics of the students' (transformed) sample data. The key difference here is that we are assuming the population slope is equal to zero.

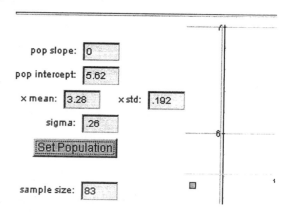

Sampling Regression Lines

(d) Set the sample size to be 83 and click the Draw Samples button. Did you obtain the same sample regression line as the students?

(e) Click the Draw Samples button again. Record the regression line displayed at the bottom. Did you obtain the same sample regression line as in (d)?

Discussion: These questions reveal once again the omnipresent phenomenon of *sampling variability*. Just as you have seen that the value of a sample mean varies from sample to sample, the value of a sample proportion varies from sample to sample, and the value of a Chi-Square statistic varies from sample to sample, now you see that the value of a sample regression line also varies from sample to sample. Once again we can use simulation to explore the long-run pattern of this variation, this time of the *sampling distributions* of the sample slope coefficients and the sample intercept coefficients.

(f) Change the "num samples" box from 1 to 100. Click the Draw Samples button. As the samples scroll by, describe the pattern of variation that you see in the simulated regression lines

(g) A new window pops up showing a dotplot of the sample slopes and a dotplot of the sample intercepts. Record the shape, center, and spread of each distribution in the table below.

	Distribution of b_0	Distribution of b_1
shape		
mean		
standard deviation		

Are the *means* of these distributions roughly what you expected? Explain.

(h) Click Reset. Change the value of sigma from .26 to .13 and click Set Population. How does this change the scatterplot?

(i) How do you think this will change the behavior of the distribution of sample slopes and the distribution of the sample intercepts? (shape, center, spread)

(j) Click the Draw Samples button. Was your conjecture in (i) correct? (You might want to look at both dotplot windows and the table in (g).)

(k) Change the value of sigma back to .26 (and click Set Population) but change the value of SD(X) from .192 to .1. Click the Set Population button. How does this change the scatterplot?

(l) How do you think this will change the behavior of the distribution of sample slopes and the distribution of the sample intercepts?

(m) Click the Draw Samples button. Was your conjecture in (l) correct?

(n) Change the value of SD(X) back to .192 (Set Population) and change the sample size from 83 to 40. Conjecture what will happen to the sampling distributions before you click Draw Samples. Was your prediction correct?

(o) Write a paragraph summarizing how each of these quantities affects the sampling distribution of the sample slope:
- the variability about the regression line, σ
- the variability in the explanatory variable, SD(X)
- the sample size, n.